School Math Fundamentals

A. B. Schenk

ISBN 978-0-9810119-0-5

Layout and formatting by
Standard Marketing & Graphics
www.standardgraphics.net

Front cover photo from Matton Images, LLC

Contents

Chapter 1

Multiplication and Division

Section 1.1

The Real Number Line

This book deals only with the set of real numbers which are numbers that are either rational or irrational. A rational number is a number that can be put in the form $\frac{a}{b}$ (a divided by b) where a and b are integers and b is not equal to 0 ($b \neq 0$). Integers are numbers such as {…-3,-2,-1, 0, 1, 2, 3…}. The ellipsis "…" means "and so on". An irrational number is a number that cannot be put in the form $\frac{a}{b}$.

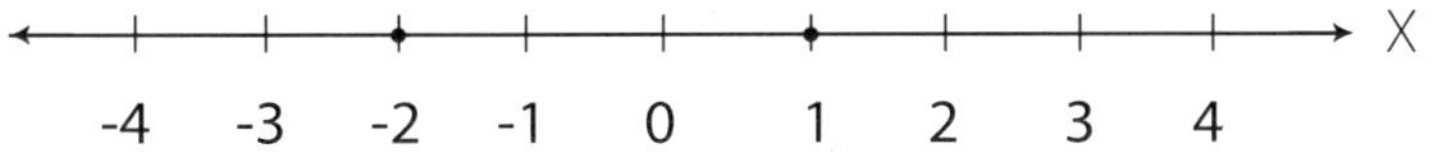

The real number line is a line that extends infinitely in both the left and right directions and has an origin corresponding to the number 0. Numbers to the right of the origin are positive (+) and numbers to the left of the origin are negative (-).

Every real number corresponds to one and only one point on the real number line. For example, the number 1 corresponds to the point one unit to the right of the origin and the number -2 corresponds to the point two units to the left of the origin. An irrational number such as $\sqrt{2}$ also corresponds to a single unique point on the real number line (A proof that $\sqrt{2}$ is irrational is given in Appendix A).

Section 1.2

Basics of Multiplication

Let's start by looking at 8 x 6. Here 8 is the multiplicand and 6 is the multiplier. We can think of this operation as taking 6 sets where each set contains 8 items. We can also think of this operation as 6 times 8, 6 of 8, six eights, etc.

This single row of eight apples represents a single set containing eight apples.

If we add five more rows of apples, we have a total of six rows, or six sets of eight apples as shown below.

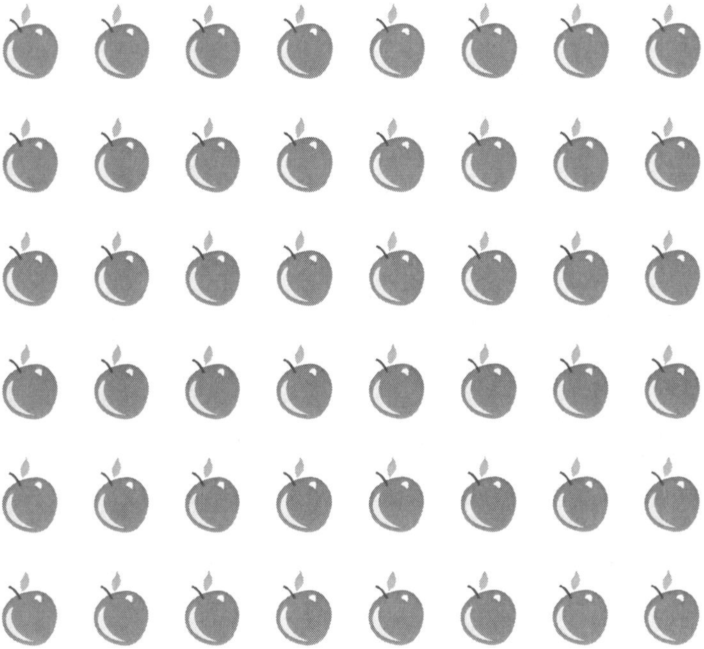

The number of apples in a single row is the multiplicand and the number of rows is the multiplier. The six rows of eight apples represents multiplying 8 by 6 and is equal to 48 apples.

If we rotate this array of apples 90 degrees, we now have a new array containing eight rows where each row contains six apples.

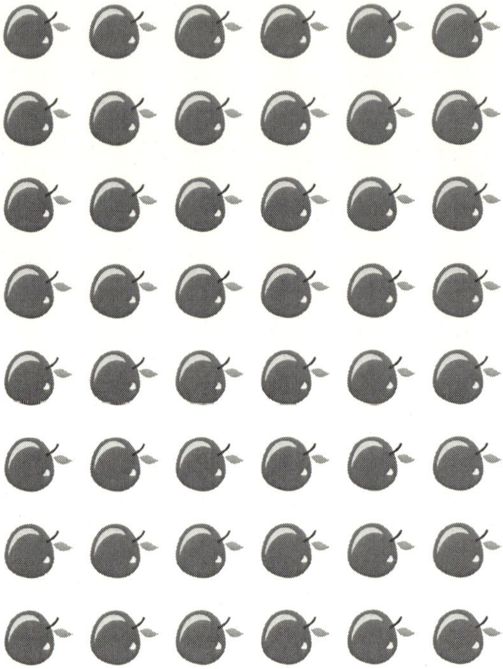

In other words, we now have eight sets of six apples, or 6 x 8, and the multiplicand in this case is 6 and the multiplier is 8. Since the total number of apples in the array has not changed, 6 x 8 = 48.

This example illustrates the commutative property of multiplication for the set of natural numbers. If a and b are natural numbers, then $a \times b = b \times a$. That is, b sets of a is equal to a sets of b. The natural numbers are a subset of the set of integers and are just the integers greater than or equal to 1, i.e. the numbers 1, 2, 3, and so on.

(1.1) $a \cdot b = b \cdot a$

Here the • means the same as the multiplication sign x and we will use them interchangeably.

From the illustration below we can see that six sets (or six rows) of eight apples is equal to four sets of eight apples plus two sets of eight apples, or in equation form,

$8 \cdot 6 = 8 \cdot (4 + 2) = 8 \cdot 4 + 8 \cdot 2.$

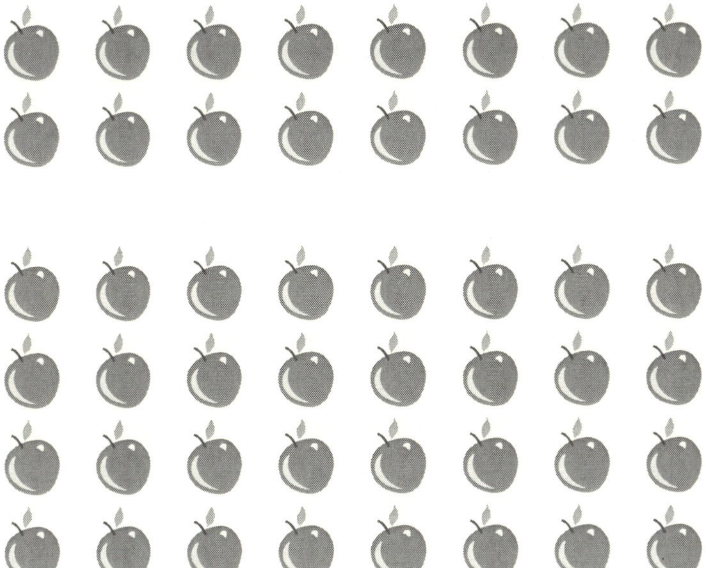

This illustrates the distributive property of multiplication over addition for the set of natural numbers. That is, (*b* plus *c*) sets of *a* is equal to *b* sets of *a* plus *c* sets of *a*, or written in equation form,

(1.2) $a \cdot (b + c) = a \cdot b + a \cdot c.$

The commutative property of multiplication (1.1) and the distributive property of multiplication over addition (1.2) should mean $(b + c) \cdot a = a \cdot b + a \cdot c$, or equivalently $(b + c) \cdot a = b \cdot a + c \cdot a$, and if you look at one of the arrays, you can see that this is in fact true.

Remember that the multiplicand is to the left of the multiplication sign and the multiplier is to the right so the variables that we use are not what's important – it's the position that's important.

(1.3) $$(a + b) \bullet c = a \bullet c + b \bullet c.$$

The expression $(a + b) \bullet (c + d)$ can be solved using both (1.2) and (1.3) as follows.

$(a + b) \bullet (c + d)$

First apply (1.2) to get

$= (a + b) \bullet c + (a + b) \bullet d$

...and then apply (1.3) twice to get

$= a \bullet c + b \bullet c + a \bullet d + b \bullet d.$

Therefore, we can see that

$(a + b) \bullet (c + d) = a \bullet c + b \bullet c + a \bullet d + b \bullet d$

or equivalently

$$(1.4) \qquad (a + b) \cdot (c + d) = a \cdot c + a \cdot d + b \cdot c + b \cdot d.$$

(Here we have used the commutative property of addition, $a + b = b + a$, to interchange the positions of $a \cdot d$ and $b \cdot c$.)

Now suppose that instead of apples in the rectangular array we have baskets. That is, each row contains eight baskets and there are a total of six rows. If we had, say, 3 apples in each basket, then the total number of apples would be 3 apples per basket times (8 x 6 baskets) = 144 apples. We would get this same number of apples if we multiply the 3 apples per basket by the 8 baskets per row to get the number of apples per row and then multiply this number by 6 rows. Here we see that $3 \cdot (8 \cdot 6) = (3 \cdot 8) \cdot 6$ and this example illustrates the Associative Property of Multiplication for the set of natural numbers. That is, for natural numbers a, b, and c,

$$(1.5) \qquad a \cdot (b \cdot c) = (a \cdot b) \cdot c.$$

Section 1.3

Negative Numbers

One way that we can look at the operation of adding 2 plus 3 $(2 + 3)$ is to think of it as starting out at the position 2 on the real number line and then moving to the right 3 more units to arrive at the position 5 units to the right of the origin.

Now, if, for example, we want the equation $2 - 3$ to have a solution, we must expand the number system that we have been using so far (which has been only the set of positive integers) to introduce the idea of a negative number.

First, we define $a - b$ to be equal to $a + (-b)$ for all positive integers a and b.

Therefore, $2 - 3$ would be equal to $2 + (-3)$ and we can represent this operation by starting out at the position 2 on the real number line as before but since we are adding the negative number -3 we now move left 3 units and we see that we arrive at the point one unit to the left of the origin, which as we saw earlier corresponds to the number -1. Therefore $2 - 3 = -1$.

Multiplication Involving Negative Numbers

Now, how should we define multiplication between positive and negative numbers? We know that $1 + (-1) = 0$ so let's look at the equation

$$1 \cdot 0 = 0$$

or equivalently

$$1 \cdot [1 + (-1)] = 0.$$

We have seen that

$$a \cdot (b + c) = a \cdot b + a \cdot c$$

is true for the set of natural numbers and if we want this axiom to be preserved with our new expanded number system, which now includes negative integers, then we must have

$$1 \cdot [1 + (-1)]$$

$$= 1 \cdot 1 + 1 \cdot (-1)$$

$$= 1 + 1 \cdot (-1)$$

and since we know that this must equal zero we have

$$1 + 1 \cdot (-1) = 0.$$

Therefore, in order for (1.2) to remain true now that we have expanded our number system from the set of positive integers to include the negative integers, we must *define* $1 \cdot (-1)$ to be equal to -1.

Similarly,

$0 \cdot 1$ must equal 0 so (1.3) gives

$$[1 + (-1)] \cdot 1$$

$$= 1 \cdot 1 + (-1) \cdot 1$$

$$= 1 + (-1) \cdot 1$$

and since $1 + (-1) \cdot 1$ must equal 0, $(-1) \cdot 1$ must be equal to -1.

We *define* a negative number multiplied by a positive number to be a negative number in order to preserve the axioms which were true for the set of positive integers.

We saw that a consequence of the commutative property of multiplication and the distributive property of multiplication over addition for the natural numbers is that

$$(a+b)\cdot(c+d) = a\cdot c + a\cdot d + b\cdot c + b\cdot d \qquad (1.4)$$

so let's look at

$$0\cdot 0 = [1+(-1)]\cdot[1+(-1)] = 0 \quad .$$

If we want the axioms that were true for the set of natural numbers to be preserved with our new expanded number system which now includes the negative integers, then we must have

$$= [1+(-1)]\cdot[1+(-1)] = 0$$

$$= 1\cdot 1 + 1\cdot(-1) + (-1)\cdot 1 + (-1)\cdot(-1) = 0 \quad .$$

And since we already saw that $1\cdot(-1)$ and $(-1)\cdot 1$ must each be equal to -1 we get

$$= 1 + (-1) + (-1) + (-1)\cdot(-1) = 0$$

$$= -1 + (-1)\cdot(-1) = 0$$

Therefore, in order for (1.4) to remain true now that we have expanded our number system from the set of positive integers to include the negative integers, we must *define* $(-1)\cdot(-1)$ to be equal to $+1$.

Section 1.4

Division Involving Integers

As was pointed out previously, in multiplication we have a multiplicand and a multiplier and you can think of the multiplicand as a set with a certain number of elements in it and the multiplier as the number of sets.

Example: (3 apples per person) x (5 people) gives us a total of 15 apples. Here the multiplicand is the 3 apples and the multiplier, or sets, are the five people.

We can look at the opposite of this operation two ways.

15 apples / (3 apples per person) = 5 people

or

15 apples / (5 people) = 3 apples per person.

In the first case we are asking "How many <u>sets</u> (people in this case) containing 3 apples do we have to take to get a total of 15 apples?" and in the second case we are asking "How many <u>apples</u> (per person) do we need so that when we take 5 sets of them we have a total of 15 apples?"

In reality we would think of the multiplier not as a

physical object like a basket or a person but rather as a "pure" number. 15 apples / 3 apples would be equal to the pure number 5. On the other hand, 15 apples divided by the pure number 5 would be equal to 3 apples. In the first case, 15 apples / 3 apples, you can think of the apples as the units in both the numerator and denominator which cancel giving us the unit-less pure number 5 and in the second case, 15 apples / 5 equals 3 apples because the 5 in the denominator is unit-less and therefore it does not cancel out the units of apples is the numerator.

Axioms

The following axioms are true for all integers a, b and c.

Closure property under Addition

$a+b$ is an integer

Closure Property under Multiplication

$a \bullet b$ is an integer

Commutative Property of Addition

$a + b = b + a$

Commutative Property of Multiplication

$a \bullet b = b \bullet a$

Associative Property of Addition

$(a + b) + c = a + (b + c)$

Associative Property of Multiplication

$(a \cdot b) \cdot c = a \cdot (b \cdot c)$

Distributivc Property of Multiplication over Addition

$a(b + c) = ab + ac$

Additive Identity

$a + 0 = a$

Property of Additive Inverse

$a + (-a) = 0$

Multiplicative Identity

$a \cdot 1 = a$

Section 1.5

Multiplication of Rational Numbers

As was stated earlier, a rational number is a number of the form $\frac{a}{b}$ where a and b are integers and b is not equal to zero $(b \neq 0)$. For example, $\frac{1}{5}$ means 1 divided by 5.

Now suppose that we have a loaf of unsliced bread and we ask "How much of the loaf, when multiplied by five equals the whole loaf?" Obviously the answer is one-fifth of the loaf. That is, one part out of five equal parts of the whole loaf. Now if we take that $\frac{1}{5}$ of a loaf and multiply it first by 5, we get five-fifths, or one whole loaf, and then if we multiply again by 2 we get ten-fifths, or two whole loaves. Now, if we instead first multiply the one-fifth of the loaf by 2 to get two-fifths of a loaf and then multiply by 5 we still get the same amount of two whole loaves (because $\frac{1}{5} \cdot 5 \cdot 2 = \frac{1}{5} \cdot 2 \cdot 5$ by the commutative property of multiplication). And finally, if we reverse this last step i.e., take the two whole loaves of bread and divide by 5, we get back to two-fifths of a loaf of bread. Therefore we see that two loaves of bread divided by 5 is equal to two-fifths (2 parts out of five equal parts) of one whole loaf. And in general, $\frac{a}{b}$ or *a* divided by *b* is equal to *a* parts out of *b* equal parts of one whole.

It's obvious that one-fifth multiplied by two is equal to

two-fifths, or $\frac{2}{5}$, and this is why we define $\frac{1}{5}$ multiplied by 2 to be equal to $\frac{2}{5}$. And in general, we define $\frac{a}{b} \times c$ to be equal to $\frac{a \cdot c}{b}$ where *a, b* and *c* are integers $(b \neq 0)$.

(1.6) Definition: $\frac{a}{b} \cdot c = \frac{a \cdot c}{b}$

Section 1.6

Division of Rational Numbers

How should we define division with rational numbers? For example, what should we define the result of $\frac{\frac{1}{5}}{3}$ to be? If we start again with one loaf of bread and then divide by 5, we get $\frac{1}{5}$ (one-fifth) of a loaf of bread. If we then divide this $\frac{1}{5}$ by 3 we get a value, *x*, that when multiplied by 3 gets us back to $\frac{1}{5}$ and then when multiplied again by 5 gets us back to one whole loaf. Clearly this means that $x = \frac{1}{15}$ because $(x \cdot 3) \cdot 5 = x \cdot (3 \cdot 5) = x \cdot 15 = 1$. That is, *x*, when multiplied by 15 = 1.

(1.7) Definition: $\frac{\frac{a}{b}}{c} = \frac{a}{b \cdot c}$

Now suppose that we want to know what two-thirds of four-fifths is. I.e., we want to multiply $\frac{4}{5}$ by $\frac{2}{3}$.

If we divide $\frac{4}{5}$ by 3, we will find what one-third of four-fifths is and then if we multiply this new amount by 2, we will find what two-thirds of four-fifths is. Dividing $\frac{4}{5}$ by 3 we get $\frac{4}{15}$ and then multiplying $\frac{4}{15}$ by 2 we get $\frac{8}{15}$. Therefore two-thirds of four-fifths is equal to $\frac{4}{5} \cdot \frac{2}{3} = \frac{8}{15}$.

This example illustrates how we define the multiplication of two rational numbers.

(1.8) $\frac{a}{b} \cdot \frac{c}{d} = \frac{a \cdot c}{b \cdot d}$

Finally let's see how we can divide one rational number by another rational number.

Let's look at the example of dividing $\frac{1}{2}$ by $\frac{2}{3}$.

$$\frac{\frac{1}{2}}{\frac{2}{3}}$$

Here we are asking the question "What do we have to multiply $\frac{2}{3}$ by to get $\frac{1}{2}$?"

Using (1.8) we can see that if we first multiply $\frac{2}{3}$ by $\frac{3}{2}$ we get $\frac{2}{3} \cdot \frac{3}{2} = \frac{2 \cdot 3}{3 \cdot 2} = 1$. Now to get $\frac{1}{2}$ we simply multiply again by $\frac{1}{2}$. Since we first multiplied $\frac{2}{3}$ by $\frac{3}{2}$ and then by $\frac{1}{2}$ we have effectively multiplied it by a total of $\frac{3}{2} \cdot \frac{1}{2} = \frac{3}{4}$ and therefore $\frac{\frac{1}{2}}{\frac{2}{3}} = \frac{3}{4}$.

This illustrates how we define the division of two rational numbers.

(1.9) $\frac{\frac{a}{b}}{\frac{c}{d}} = \left(\frac{d}{c}\right) \cdot \left(\frac{a}{b}\right) = \frac{d \cdot a}{c \cdot b}$

Section 1.7

Irrational Numbers

We saw that the natural numbers are the set of numbers {1,2,3,…} and that the integers are the set of numbers {…-3,-2,-1,0,1,2,3,…}. We also saw that the set of rational numbers are numbers of the form $\frac{a}{b}$ where *a* and *b* are integers and *b* is not equal to 0. There is another set of numbers, however, that don't fit any of these sets. They are the set of irrational numbers. An irrational number is a number that cannot be put into the form $\frac{a}{b}$ where *a* and *b* are integers. One such irrational number is the square root of 2 or $\sqrt{2}$ and it is the number that when multiplied by itself equals 2. I.e., $\sqrt{2} \cdot \sqrt{2} = 2$.

The union of the set of rational numbers and the set of irrational numbers make up the entire set of real numbers. The union of two sets is the set made up of elements that are in either the first set or the second set. For example if we define Set A to be {1,2,3} and Set B to be {7,8,9},

then the union of these two sets is $\{1,2,3,7,8,9\}$.
The intersection of two sets A and B is the set made up of elements that are in both set A and set B. For the sets A and B given above, this is the empty set (the set containing no elements) because there is no element that is in both set A and set B. The intersection of the sets of rational and irrational numbers is also the empty set because if a number is rational, it is not irrational and vice versa.

Section 1.8

Axioms of the Real Numbers

The following axioms are true for all *real* numbers x, y and z.

Closure property under Addition

$x+y$ is a real number

Closure Property under Multiplication

$x \cdot y$ is a real number

Commutative Property of Addition

$x + y = y + x$

Commutative Property of Multiplication

$x \bullet y = y \bullet x$

Associative Property of Addition

$(x + y) + z = x + (y + z)$

Associative Property of Multiplication

$(x \bullet y) \bullet z = x \bullet (y \bullet z)$

Distributive Property of Multiplication over Addition

$x (y + z) = xy + xz$

Additive Identity

There exists a unique real number 0 such that

$x + 0 = x$

Multiplicative Identity

There exists a unique real number 1 such that

$x \bullet 1 = x$

Property of Additive Inverse

For every real number x there exists a unique real number $-x$ such that

$x + (-x) = 0$

Multiplicative Inverse

For every real number $x \neq 0$, there exists a unique real number $\frac{1}{x}$ such that

$x \cdot \frac{1}{x} = 1$

Chapter 2
Basic Math Operations

Section 2.1

Positional Number System

In the positional number system that we use, the first position to the left of the decimal point represents the set of ones and the second position to the left of the decimal point represents the set of tens and the third position represents the set of hundreds and so on. Each position represents a set that is ten times more than the position to its immediate right. For example, the number 519 represents five sets of 100 plus one sets of ten plus nine ones.

Adding Numbers

When we add a series of numbers we write the numbers one above another as shown.

$$\begin{array}{r} 519 \\ 227 \\ 89 \\ \underline{12} \\ 847 \end{array}$$

Then we add up the numbers in the right-most column (in this case the ones column) and if its sum is greater than ten, we add the result's corresponding set of tens to the column to the left and its ones are written at the bottom of the ones column. We then move one column to the left to the tens column and add up the sets of tens (including the sets of ten carried over from adding up the ones in the ones column) and if its sum is greater than ten, we add the result's corresponding set of hundreds to the column to the left and its tens are written at the bottom of the tens column. We keep repeating this procedure until we run out of columns and at this point the final sum is the number we have written at the bottom of the list of numbers.

Section 2.2

Multiplication Algorithm

If we take the number 234 and add a zero in the right-most position to get 2340 we have 234 tens (sets of ten) instead of the 234 ones that we had previously. We can multiply any number by ten simply by adding a zero to the right-most position and we can multiply a number by 100 by adding two zeros in the right-most two positions and so on. If we want to multiply the number by say, twenty, then we could first multiply it by ten (by putting a zero in the right most position) and then multiplying it by two.

For example, $(234) \cdot 20$

$= (234) \cdot (2 \cdot 10)$

$= (234) \cdot (10 \cdot 2)$

$= (234 \cdot 10) \cdot 2$

$= (2340) \cdot 2$

$= 2000 \cdot 2 + 300 \cdot 2 + 40 \cdot 2 + 0 \cdot 2$

$= 4000 + 600 + 80 + 0$

$= 4680.$

Another example: $(234) \cdot 27$

$= (234) \cdot (20 + 7)$

$= (234) \cdot 20 + (234) \cdot 7$

$= 4680 + (234) \cdot 7$

$= 4680 + (200 \cdot 7 + 30 \cdot 7 + 4 \cdot 7)$

$= 4680 + (1400 + 210 + 28)$

$= 6318.$

Section 2.3

Division Algorithm

Suppose that we want to divide 768 by 3.

$$3\overline{)\ 768}$$

The first thing we do is ask "How many sets of threes can we take away from 768 without taking away more than 768?"

If we place a 2 above the 7, this represents 200 because the 7 is in the hundreds column. Two hundred sets of 3 equals 600 so we subtract 600 from 768 and we are left with 168.

```
   2
3) 768
  -600
  ----
   168
```

(Note that if we would have placed a 3 above the 7, this would represent 300 and 300 sets of 3's equals 900 which is more than the 768 we started with.)

So now we ask "How many sets of threes can we take away from the remaining 168 without taking away more than 168?"

If we place a 5 above the number 6 in 768, this represents 50 because it is in the tens column. Fifty sets of 3's is equal to 150 and if we now subtract 150 from the 168, we are left with 18.

```
     25
   ____
3) 768
  -600
  ----
   168
  -150
  ----
    18
```

Now we ask "How many sets of threes can we take away from the remaining 18 without taking away more than 18?"

If we place a 6 above the number 8 in 768, this simply represents 6 because it is in the ones column. Six sets of 3's is equal to 18 and we now subtract the 18 from the remaining 18 and are left with 0.

```
    256
   ____
3) 768
  -600
  ----
   168
  -150
  ----
    18
   -18
   ---
     0
```

Therefore we see that, 200 sets of 3 plus 50 sets of 3 plus 6 sets of three equal 768. That is, 256 sets of 3 equal 768.

If we would have divided 769 by 3, we would have followed the same procedure but gotten a remainder of one.

```
     256
   -----
3) 769
  -600
  -----
   169
  -150
  -----
    19
   -18
   ----
     1
```

That is, $769 = 3 \cdot 256 + 1$.

If we both multiply and divide the remainder,1, by 3 (so that its value is not changed), then we can rewrite this as

$$769 = 3 \cdot 256 + 1 \cdot \left(\frac{3}{3}\right)$$

$$= 3 \cdot 256 + 3 \cdot \left(\frac{1}{3}\right)$$

And if we factor out the 3, we get

$$769 = 3 \cdot [256 + \left(\frac{1}{3}\right)] .$$

That is, 769 divided by 3 is equal to

$$[256 + \left(\frac{1}{3}\right)]$$

$$= \left[256 \cdot \left(\frac{3}{3}\right) + \left(\frac{1}{3}\right)\right]$$

$$= \left[\left(\frac{768}{3}\right) + \left(\frac{1}{3}\right)\right] = \left(\frac{769}{3}\right) .$$

Section 2.4

Decimal Numbers

A decimal number is just a rational number where its denominator is some power of ten. For example, 0.1 has the number 1 in the tenths column so it is equal to 1 divided by ten.

For the decimal number 0.27 the two is in the tenths column and the seven is in the hundredths column. Therefore, we have two tenths plus seven hundredths. Since two-tenths are equal to twenty-hundredths we can write this total as twenty seven hundredths or 27/100.

Multiplying Decimal Numbers

In section 2.2 we used the multiplication algorithm to multiply 234 by 27 to get 6318 so let's see what happens when we multiply 2.34 by 2.7.

2.34 is equal to $\frac{234}{100}$ and 2.7 is equal to $\frac{27}{10}$ so we are multiplying $\frac{234}{100}$ by $\frac{27}{10}$.

We saw in section 1.6 that multiplication involving rational numbers, or fractions, is

defined by $\frac{a}{b} \cdot \frac{c}{d} = \frac{a \cdot c}{b \cdot d}$

so

$$\frac{234}{100} \cdot \frac{27}{10} = \frac{234 \cdot 27}{100 \cdot 10}.$$

We see that the denominator must equal 1000 and since we already used the multiplication algorithm to find that $234 \cdot 27 = 6318$ we have

$$\frac{234 \cdot 27}{100 \cdot 10} = \frac{6318}{1000}.$$

Now, to divide 6318 by 1000 we simply move the decimal point three places to the left to get 6.318.

So when we multiply two decimal numbers together such as $(2.34) \cdot (2.7)$ we can initially disregard the decimal points and use the multiplication algorithm to find the total as if the two numbers weren't fractions and then we count up the number of positions in each number that are to the right of the decimal point and add them and finally we move the decimal point that total number of positions to the left for the number we just calculated.

Decimal Fractions

We can convert a rational number like ¼ to a decimal number using an algorithm almost identical to the division algorithm.

First, place the decimal point to the right of the 1 and add a zero to the right of the decimal point. Now we can think of the 1 as ten tenths. Now follow the step by step procedure of the division algorithm.

$4\overline{)\,1.0}$

Four goes into ten tenths two tenths times so place a 2 at the top of the tenths column, and subtract 8/10 leaving 2/10.

$$\begin{array}{r} 0.2 \\ 4\overline{)\,1.0} \\ -0.8 \\ \hline 0.2 \end{array}$$

2/10 is equal to twenty hundredths so we can add a zero the right of the 2 in the hundredths column.

$$\begin{array}{r} 0.2 \\ 4\overline{)\,1.0} \\ -0.8 \\ \hline 0.20 \end{array}$$

Four goes into the remaining twenty hundredths five hundredths times so place a five at the top of the hundredths column. Subtracting the five-hundredths times 4, or twenty-hundredths, leaves us with the remainder of zero.

$$\begin{array}{r} 0.25 \\ 4\overline{)\,1.0} \\ -0.8 \\ \hline 0.20 \\ -0.20 \\ \hline 0 \end{array}$$

We've therefore shown that 1 divided by 4 is two-tenths (or twenty-hundredths) plus five-hundredths equal to twenty five hundredths or 0.25.

Non-terminating, Repeating Decimal Numbers

Now let's look at some non-terminating decimal numbers that have digits that repeat in some manner.

0.3333… where the 3's repeat endlessly is a rational number. We can show this by letting $x = 0.3\ldots$ and therefore $10x = 3.3\ldots$.

$10x - x = 3.3\ldots - 0.3\ldots$

$10x - x = 9x$ and $3.3\ldots - 0.3\ldots = 3$ therefore we have $9x = 3$.

Dividing both sides of $9x = 3$ by 9 we get $x = \frac{3}{9} = \frac{1}{3}$.

As another example consider 5.65749749…
where the group of digits 749 repeats endlessly.
If we let $x = 5.65749749\ldots$, then $100x = 565.749\ldots$ and $100{,}000x = 565749.749\ldots$ and subtracting $100x$ from $100{,}000x$ we would get

$100{,}000x - 100x = 565749.749\ldots - 565.749\ldots$

or

$$99{,}900x = 565184$$

and dividing both sides of the equation by 99,900 we get

$$x = \frac{565184}{99{,}900} \; .$$

We already knew that a terminating decimal number is a rational number and now we can see that a non-terminating, *repeating* decimal number is also rational. A non-terminating, non-repeating decimal number, however, is irrational.

Section 2.5

Equations

Suppose that your friend has two quarters and 4 dimes for a total of 90¢ and suppose that you have two quarters, six nickels and ten pennies also for a total of 90¢. Now suppose that you and your friend arranged your coins on the table as below.

=

The amount that you and your friends have initially is equal and if you were both to divide your money by two, it could look like this.

=

Your friend now has one quarter and two dimes for a total of 45¢ and you now have one quarter, three nickels and five pennies for a total also of 45¢. If your friend now subtracts 10¢ from his coins by taking away a dime and you subtract 10¢ from your coins by taking away one nickel and five pennies, then you both now have 35¢.

=

This simple example illustrates the fact that if you have two values that are initially equal, like the two sides of an equation, and then perform the same mathematical operation (in the same sequence) on both sides of the equation, then the equality remains true even though the values may have changed.

The left side of the equation represents your friend's two quarters plus four dimes and the right side of the equation represents your two quarters, six nickels and ten pennies, and the equation remains true when you and your friend both divide your money by two and the equation still remains true when you both subtract 10¢ from your remaining coins.

Chapter 3
Proportion

Section 3.1

Direct Proportion

Suppose that you want to buy some stamps. The cost to buy two stamps is, obviously, twice the cost to buy a single stamp and the cost to buy three stamps is three times the cost to buy just one stamp. In other words the cost of buying stamps is proportional to the number of stamps purchased. Cost $\propto$ (number of stamps) with the "$\propto$" meaning "is proportionate to". We can change this statement to an equation by including a constant, k.

Thus $C = k \cdot (\text{\# of stamps})$.

If the cost to buy one stamp is 50¢, then we have

$$50¢ = k \cdot (1 \text{ stamp})$$

Therefore, $k = \frac{50¢}{\text{stamp}}$.

A value can be proportionate to more than one variable. For example, the amount of torque (or twisting force) produced by hanging a weight from a lever arm is proportionate to the weight, w, and also the distance of the weight from the pivot point.

$$\tau \propto w \cdot d$$

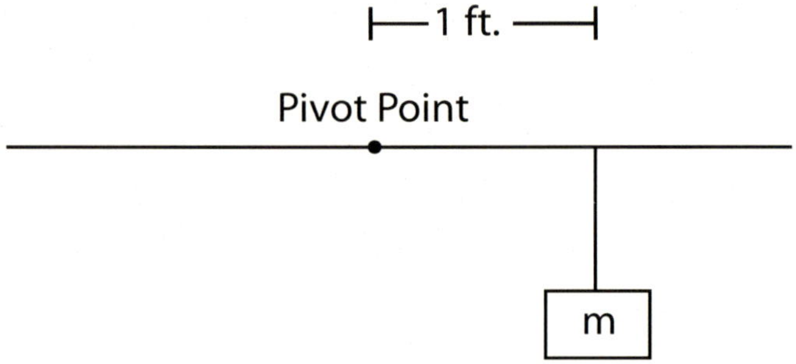

Suppose that you have a weight hanging from a point on the lever arm 1ft. to the right of the pivot point. The weight hanging from the arm will produce a torque, or twisting force, that will tend to rotate the arm clockwise. Experiment shows that if you double the distance so that the weight now hangs 2ft. to the right of the pivot point, then the same weight will now produce twice as much torque as when it was only 1ft. from the pivot point. If you were to move the weight so that it is hanging 3ft. from the pivot, then it will produce three times as much torque as when it was hanging only 1ft. from the pivot point. The distance from the pivot point is the lever arm length and we see that, for a given weight, torque is proportional to the lever arm length, *d*.

On the other hand if we keep the weight a fixed distance from the pivot point but double its weight, experiment shows that the torque is also doubled.

$$\tau \propto w \cdot d$$

We can turn the proportion into an equation by introducing a constant, *k*, so that

$$\tau = \mathrm{k} \cdot wd \; .$$

A constant is a number that is fixed or doesn't change. Suppose that the torque value is 1 when a 1 lb weight is hanging 1 ft. to the right of the pivot point. In this case $\tau = 1$, $w = 1$ and $d = 1$ so that we have

$$1 = k \cdot 1 \cdot 1$$

and in this case our constant, *k*, must be equal to 1 and therefore we have

$$\tau = w \cdot d.$$

In the equation describing torque you shouldn't think

of torque as being equal to weight multiplied by distance but rather as the magnitude of the torque being proportional to the weight and its distance from the pivot.

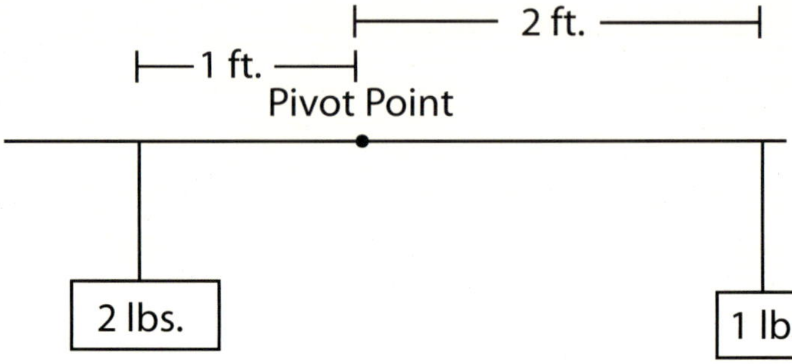

A 2 pound weight hanging from a point one foot to the left of the pivot point and a 1 pound weight hanging from a point two feet to the right of the pivot point would produce torque values that are equal in magnitude but opposite in direction so that the lever arm would not rotate at all.

Section 3.2

Inverse Proportion

There can also be an inverse relationship between two values. For example, Isaac Newton realized that the acceleration that a given force imparts on different objects is inversely proportional to the object's mass.

Newton saw, by experiment, that a force that produced an acceleration of a_1 on a 1kg mass would produce an acceleration of only $\frac{1}{2} \cdot a_1$ on a 2kg mass and an acceleration of $\frac{1}{3} \cdot a_1$ on a 3kg and so on.

Thus, for a fixed force the magnitude of acceleration is inversely proportional to mass.

$$a \propto \frac{1}{m}$$

Newton then defined force in terms of the acceleration that it imparts to a mass. He said that if one force causes a given mass to accelerate at a rate of a_1 and a second force causes the same mass to accelerate at $2a_1$, then the second force is defined to have twice the magnitude of the first force.

So, acceleration is inversely proportional to mass by experiment and directly proportional to force by definition. Combining these two gives us:

$$a \propto \frac{F}{m}$$

This can be rewritten as $F \propto m \cdot a$ or in equation form

$F = k \cdot m \cdot a$ for some constant *k*.

The force that produces an acceleration of one meter per second squared $(\frac{1m}{s^2})$ when applied to a one kilogram mass is defined to be one Newton (*N*) so

$$F = k \cdot m \cdot a$$

gives us

$$1N = k \cdot 1kg \cdot \frac{1m}{s^2}$$

and consequently *k* must have a magnitude of 1.

$$1N = 1kg \cdot \frac{1m}{s^2}$$

It is convention that we "multiply" the units of the variables together the same way that we multiply the magnitudes of the variables so that in this case we agree that the Newton has units of $kg \cdot \frac{1m}{s^2}$ (kilogram meters per second squared) so that both sides of the equation have the same units.

Section 3.3

Area

Suppose that we have a square tile that is 1foot by 1foot as shown below. We define its area to be $1ft.\times 1ft. = 1\,ft^2$ or one square foot.

If we place a second one square foot tile to the right of this first tile, then they will have a combined length of 2ft. and a width of 1ft. Obviously the two tiles together cover two square feet and the product of their length and width is now $2ft.\times 1ft. = 2\,ft^2$ or two square feet.

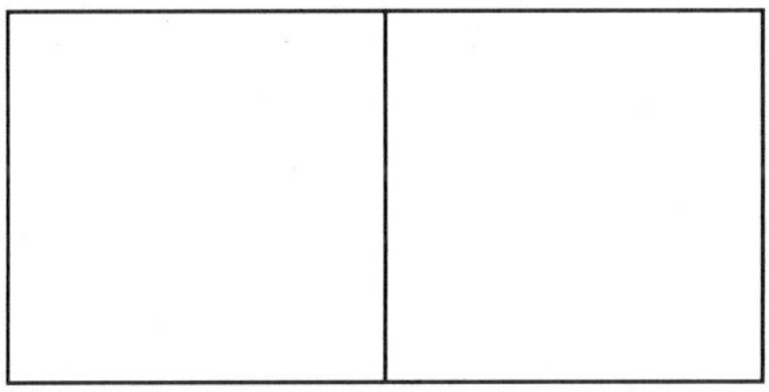

We can see that area is proportional to length. If we place four more one square foot tiles above the first two titles as shown below, then we can see that they cover a total of six square feet. And the product of their length and width is $2ft.\times 3ft. = 6\,ft^2$ or square feet.

We see that area is also proportional to width.

$$Area \propto l \cdot w$$

or in equation form

$$Area = k \cdot l \cdot w$$

Since $1\,ft^2 = k \cdot 1ft.\times 1ft.$

$$k = \frac{(1ft^2)}{1ft.\times 1ft.} = 1$$

and we have

$$Area = l \cdot w \, .$$

Section 3.4

Volume

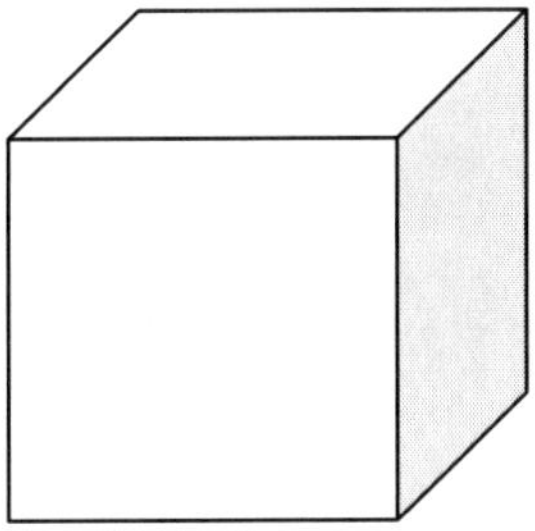

Suppose we have a cube that measures 1ft. along each side as shown above. We define this cube to have a volume of 1ft. x 1ft. x 1ft. equal to one cubic foot.

If we place a second one cubic foot cube next to it, we obviously have a total volume of two cubic feet. And the product of the two cubes' length, width and height is now 2ft. x 1ft. x 1ft. equal to two cubic feet. We can see that volume is proportional to length. That is, when we doubled the length of the first cube we got double the volume of the first cube.

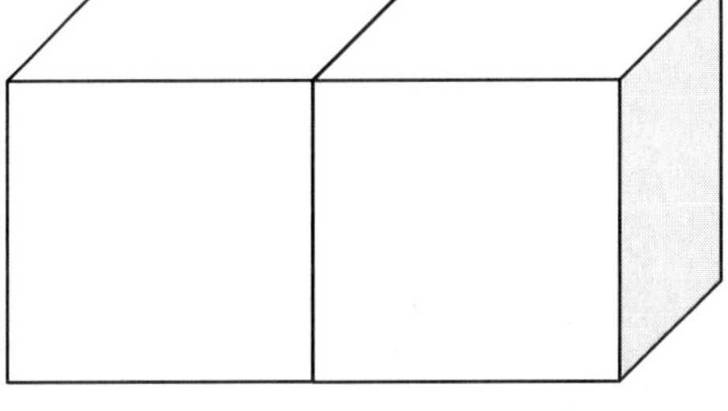

Now let us place four more one cubic foot cubes in front of the first two as shown. We obviously have a total combined volume of six cubic feet and we can see that the product of their length width and height is 2ft. x 3ft. x 1ft.= six cubic feet. We tripled the width that we had for the first two cubes and the result is that we tripled the volume. We see that volume is also proportional to width just as it is proportional to length.

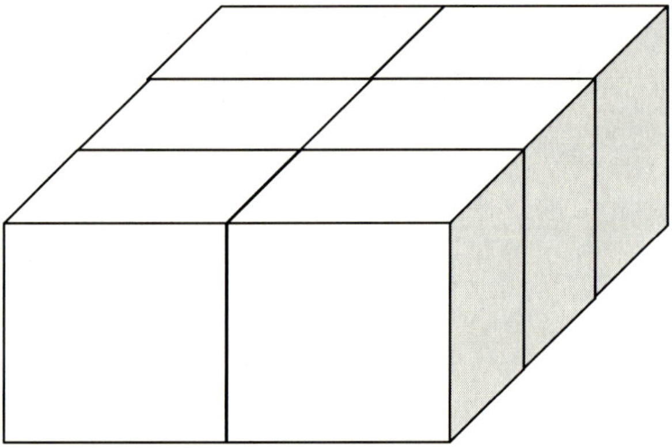

Finally, if we were to stack three more layers of cubes on top of the six cubes in the previous figure so that it is now four layers high, then we would have a total of six plus eighteen = 24 cubes for a total volume of 24 cubic feet. And the product of the new stack of cubes' length width and height is 2ft. x 3ft. x 4ft. = 24 cubic feet.

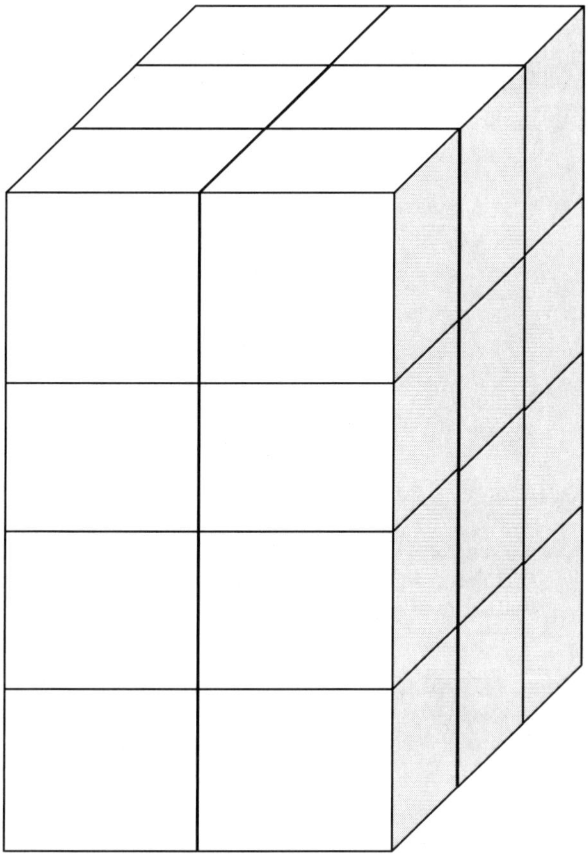

We quadrupled the height of the stack of cubes that we had previously and the result is that we quadrupled the volume. We see that volume is proportional to height just as it is proportional to the length and width.

$Volume = k \cdot l \cdot w \cdot h$ (k is the constant of proportionality)

And since $1\,ft^3 = k \cdot 1ft. \times 1ft. \times 1ft.$

$$k = \frac{(1ft^3)}{1ft. \times 1ft \times 1ft.} = 1$$

and we have

$$Volume = l \cdot w \cdot h \quad .$$

Chapter 4
The Quadratic Equation

Section 4.1

Completing the Square

When we square $(x + a)$ we get

$(x + a) \cdot (x + a)$

$= x \cdot (x + a) + a \cdot (x + a)$

$= (x \cdot x + x \cdot a) + (a \cdot x + a \cdot a)$

$= x^2 + x \cdot a + a \cdot x + a^2$

$= x^2 + 2a \cdot x + a^2.$

We can see that the coefficient of the first term, x^2, is 1 and that the coefficient of the second term, $2ax$, is $2a$ and the last term is just a^2 (which can be written as $a^2 \cdot x^0$).

So when we see an expression like

$$x^2 + 2ax + a^2$$

where the coefficient of the second term is equal to twice the value of the square root of the last term
we know that $x^2 + 2ax + a^2 = \left(x + \sqrt{a^2}\right)^2$.

This fact is useful when we want to use a technique called "completing the square" to solve the general form of the quadratic equation. Note that if the coefficient of the second term is equal to twice the value of the square root of the last term, then the last term, a^2, is equal to the square of one-half of the *coefficien*t of the second term, $2ax$.

Examples:

Suppose that we have the expression $x^2 + 8x$. What constant value would we have to add at the end of the expression in order to make it the square of x plus some constant? If we compare $x^2 + 8x$ to $(x^2 + 2ax + a^2)$, we see that $2a$ corresponds to 8 so that a would be 4 and therefore a^2 would be 16. So if we add 16 at the end of $x^2 + 8x$ to get $x^2 + 8x + 16$, it could be rewritten as

$$x^2 + 8x + 16 = (x + 4)^2 .$$

As another example, consider $x^2 + \_x + 7$. What should the coefficient of x be so that we can rewrite $x^2 + \_x + 7$ as the square of x plus some constant? If we compare $x^2 + \_x + 7$ to $(x^2 + 2ax + a^2)$ we can see that 7 corresponds to a^2 and therefore a would be equal to $\sqrt{7}$ and $2a$ would be equal to $2\sqrt{7}$. So if we insert $2\sqrt{7}$ for the coefficient of x in the expression $x^2 + \_x + 7$, we can rewrite $x^2 + 2\sqrt{7} \cdot x + 7$ as $(x + \sqrt{7})^2$.

Section 4.2

General Solution to the Quadratic Equation

The quadratic equation is an equation of the form, $ax^2 + bx + c = 0$ where a, b and c are constants and x is a variable. Constants are values that do not change while the values of variables can change (or vary).

To solve $ax^2 + bx + c = 0$ first subtract c from both sides of the equation to get

$$ax^2 + bx = -c$$

.

Then factor the constant a out of the left side of the equation.

$$a\left(x^2+\frac{b}{a}x\right)=-c$$

At this point we apply the completing the square technique on the expression inside the parentheses, $x^2+\frac{b}{a}x$.

The overall value remains unchanged if we add and subtract the square of one-half of the *coefficient* of $\frac{b}{a}x$, to the expression within the parentheses.

$$a\left(x^2+\frac{b}{a}x+\frac{b^2}{4a^2}-\frac{b^2}{4a^2}\right)=-c$$

Now we multiply out only the last term within the parentheses on the left side of the equation to get

$$a\left(x^2+\frac{b}{a}x+\frac{b^2}{4a^2}\right)-\frac{b^2}{4a}=-c$$

Now add $\frac{b^2}{4a}$ to both sides of the equation.

$$a\left(x^2+\frac{b}{a}x+\frac{b^2}{4a^2}\right)=\frac{b^2}{4a}-c$$

Since the coefficient of the second term (of the expression within the parentheses) is equal to twice the value of the square root of the last term we know that $\left(x^2 + \frac{b}{a}x + \frac{b^2}{4a^2}\right)$ can be rewritten as $\left(x + \sqrt{\frac{b^2}{4a^2}}\right)^2$ or $\left(x + \frac{b}{2a}\right)^2$.

Therefore we have

$$a\left(x + \frac{b}{2a}\right)^2 = \frac{b^2}{4a} - c$$.

Dividing both sides of the equation by *a* we get

$$\left(x + \frac{b}{2a}\right)^2 = \left(\frac{1}{a}\right) \cdot \left(\frac{b^2}{4a} - c\right)$$

or

$$\left(x + \frac{b}{2a}\right)^2 = \frac{b^2}{4a^2} - \frac{c}{a}$$.

To find a common denominator for the expression on the right side of the equation we multiply the numerator and denominator of $\frac{c}{a}$ by $4a$ to get

$$\left(x + \frac{b}{2a}\right)^2 = \frac{b^2}{4a^2} - \frac{4ac}{4a^2}$$

or

$$\left(x+\frac{b}{2a}\right)^2=\frac{b^2-4ac}{4a^2}.$$

Now taking the square root of both sides of the equation we get

$$\sqrt{\left(x+\frac{b}{2a}\right)^2}=\sqrt{\frac{b^2-4ac}{4a^2}}$$

or

$$\left(x+\frac{b}{2a}\right)=\pm\frac{\sqrt{(b^2-4ac)}}{2a}.$$

And finally, subtracting $\frac{b}{2a}$ from both sides of the equation we get

$$x=-\frac{b}{2a}\pm\frac{\sqrt{(b^2-4ac)}}{2a}$$

or

$$x=\frac{(-b\pm\sqrt{(b^2-4ac)}}{2a}.$$

Chapter 5

Trigonometric Functions

Section 5.1

Sine

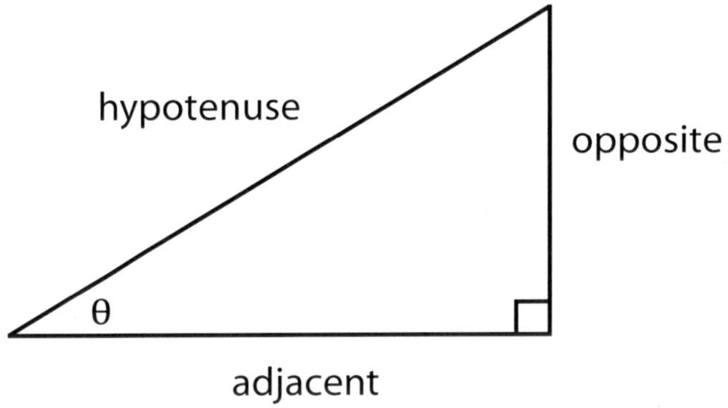

Suppose that we have a right angle triangle with the sides labeled as in the figure above.

The sine of the angle θ (θ is the Greek letter theta) is defined to be the ratio given by the length of the side of the triangle opposite the angle θ divided by the length of the hypotenuse $\sin\theta = \frac{opp}{hyp}$.

If we keep the angle θ the same but increase the lengths of all of the sides of the triangle by the same factor, the ratio $\frac{opp}{hyp}$ will still remain the same.

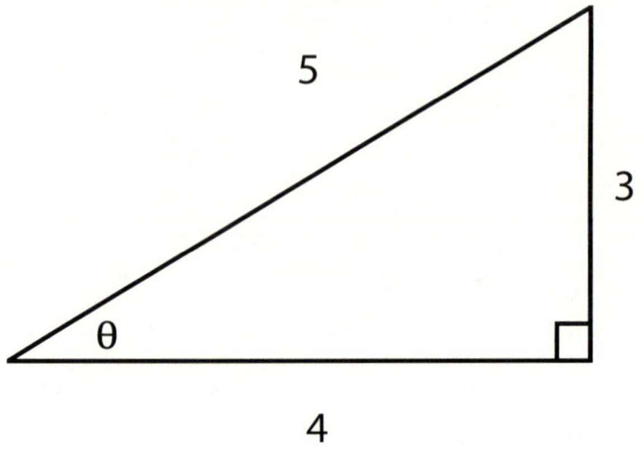

For example, if we start with the right triangle with legs of length three and four and hypotenuse of length five as shown, then sine $\theta = \frac{3}{5}$.

Now if we double the length of all of the sides of the triangle,

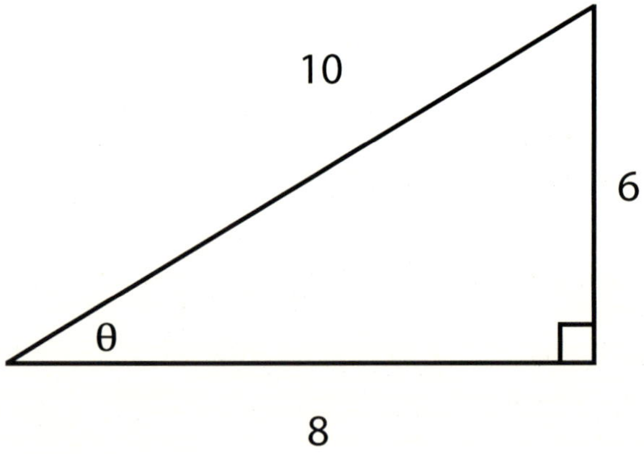

the angle θ remains the same and so does the sine of θ.

Now $\sin\theta = \frac{6}{10} = \frac{3 \cdot 2}{5 \cdot 2} = \frac{3}{5}$ which is what we had for the triangle with sides of length 3, 4 and 5.

If instead of increasing all sides of the right triangle by two, we had increased the length of all sides by the same factor, say, *a*, then the sine of θ would still be the same since we would have

$$\sin\theta = \frac{3\cdot a}{5\cdot a} = \frac{3}{5}\,.$$

Mathematicians have calculated the values of sine, cosine and tangent for angles between 0° and 360° and compiled them into tables so now when we have a right triangle where $\frac{opp}{hyp} = \frac{1}{2}$, for example, we can look up in the table to see that θ must be equal to 30°.

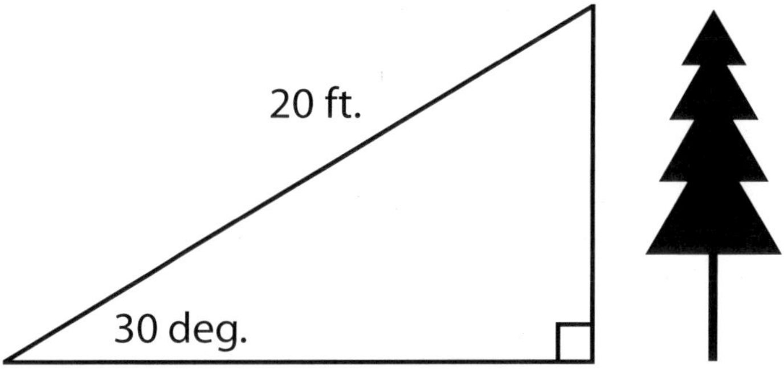

We can now also use the values of the sine function to solve practical problems. For example, if we have a 20 foot long ladder and lean it against the side of a tree as shown above and the ladder makes an angle of 30° with the ground, then we can use the values in the sine table to calculate that the height of the tree must be 10ft. because

$$\frac{height\ of\ tree}{20\ ft.} = \sin 30° = \frac{1}{2}\ .$$

Therefore the height of the tree must be equal to ½ times 20ft. or 10ft.

Section 5.2

Cosine

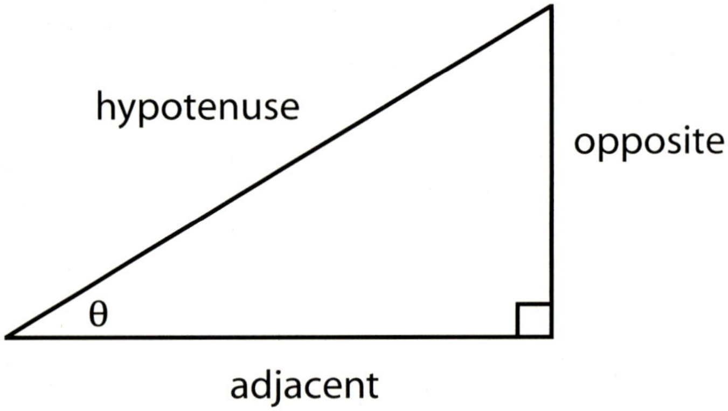

The cosine of the angle θ is defined to be the ratio given by the length of the side of the triangle adjacent the angle θ divided by the length of the hypotenuse,

$$\cos\theta = \frac{adj}{hyp}\ .$$

If we keep the angle θ the same but increase the lengths of all of the sides of the triangle by the same factor, the ratio $\frac{adj}{hyp}$ will still remain the same.

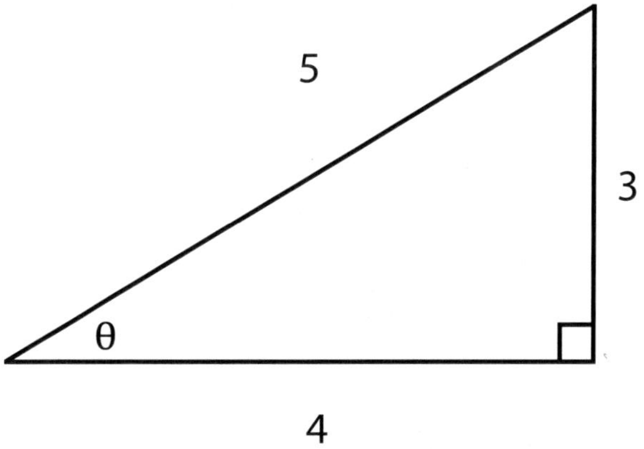

Again, if we start with the right triangle with legs of length three and four and hypotenuse of length five as shown, then cosine $\theta = \frac{4}{5}$.

Now if we double the length of all of the sides of the triangle,

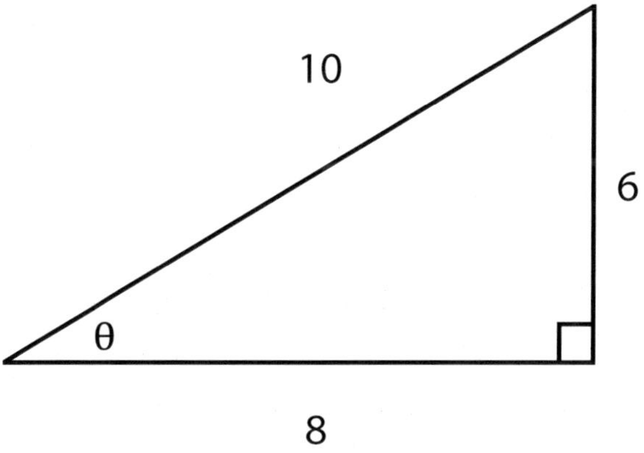

the angle θ remains the same and so does the cosine of θ since

$$\cos\theta = \frac{8}{10} = \frac{4\cdot 2}{5\cdot 2} = \frac{4}{5}$$

which is what we had for the triangle with sides of length 3, 4 and 5.

If instead of increasing all sides of the right triangle by two, we had increased the length of all sides by the same factor, a, then the cosine of θ would still be the same since we would have

$$\cos\theta = \frac{4\cdot a}{5\cdot a} = \frac{4}{5} \quad .$$

If we have a right triangle where $\frac{adj}{hyp} = \frac{1}{\sqrt{2}}$, for example, we can look up in a table of cosine values to see that θ must be equal to 45°.

And just like you could use the table of sine values to solve practical problems you can also use the table of cosine values to solve practical problems.

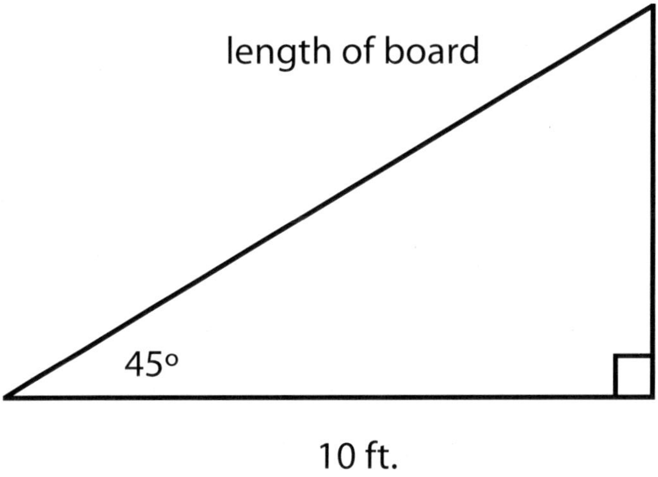

For example, if you laid a piece of wood against the wall so that the end of the piece of wood was on the ground 10ft. from the wall as shown in the figure above and the board made an angle of 45° with the ground, then

$$\frac{10\ ft.}{length\ of\ board} = \cos 45^\circ = \frac{1}{\sqrt{2}}$$

Therefore $length\ of\ board = (10\ ft.) \cdot \sqrt{2}$

Section 5.3

Tangent

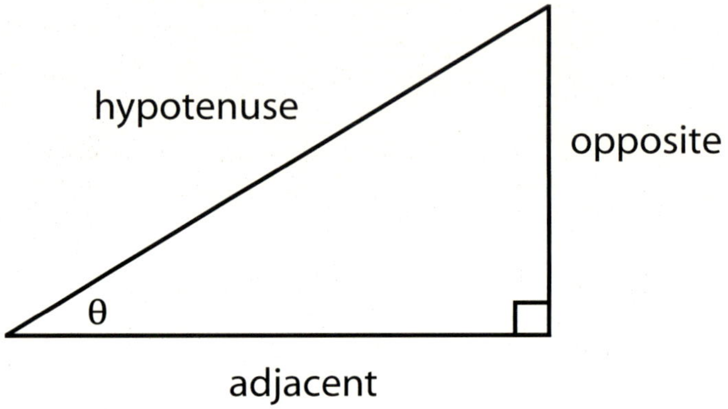

The tangent of the angle θ is defined to be the ratio given by the length of the side of the triangle opposite the angle θ divided by the length of the side of the triangle adjacent to the angle θ or $\tan\theta = \frac{opp}{adj}$.

If we keep the angle θ the same but increase the lengths of all of the sides of the triangle by the same factor, the ratio $\frac{opp}{adj}$ will still remain the same.

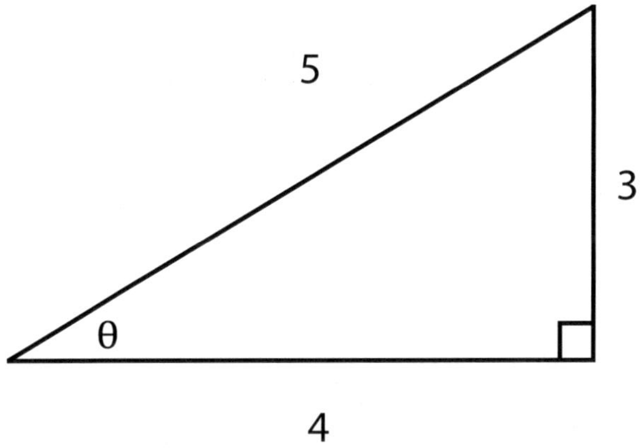

For example, if we start with the right triangle with legs of length three and four and hypotenuse of length five as shown, then tan $\theta = \frac{3}{4}$.

If we increased the length of all sides of the triangle by the same factor, *a*, then the tangent of θ would still be the same since we would have

$$\tan\theta = \frac{3 \cdot a}{4 \cdot a} = \frac{3}{4}.$$

We can now use the values of the tangent function to solve practical problems.

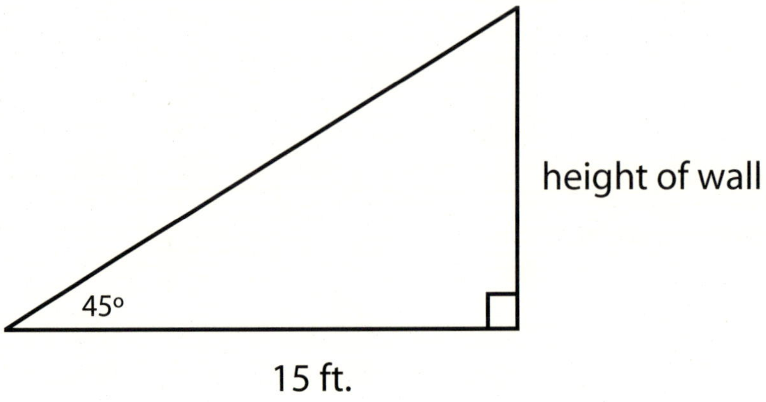

For example, if we have a length of board leaning against a wall as shown above so that it makes an angle of 45° with the ground and we also know that the board rests on the ground at a point 15 ft. from the wall, then we can use the tangent values to calculate that the height of the wall must be 15 ft. because

$$\frac{height\ of\ wall}{15\ ft.} = \tan 45° = 1\ .$$

Therefore the height of the wall must be equal to 1 times 15ft = 15 ft.

Sine, Cosine and Tangent Functions

There is the second way to define the sine, cosine and tangent functions.

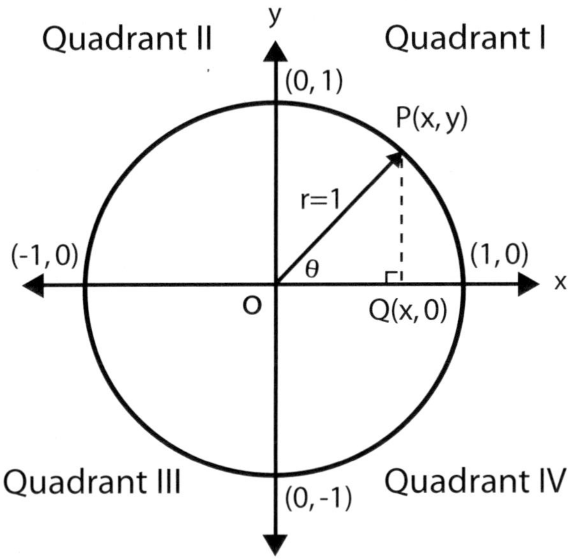

The figure above shows a circle of radius 1 that has its center at the origin of the x-y coordinate plane and a right angle triangle is inscribed within the circle.

If we start with the line segment, OP, that extends from the origin to the point (1,0) on the unit circle and rotate it counterclockwise until the point on the circle is at (x,y) as shown, then OP forms an angle θ with the positive half of the x-axis.

We now have a right triangle formed by the three points O, P and Q inscribed within the unit circle.

Using the definitions of $\sin\theta$, $\cos\theta$ and $\tan\theta$ that we saw earlier gives us:

$$\sin\theta = \frac{opp.}{hyp.} = \frac{y}{r} = \frac{y}{1} = y$$

because the length of the side PQ = y and r = 1.

Similarly

$$\cos\theta = \frac{adj.}{hyp.} = \frac{x}{r} = \frac{x}{1} = x$$

because the length of the side OQ = x and r = 1.

Finally,

$$\tan\theta = \frac{opp.}{adj.} = \frac{y}{x}.$$

And since $\sin\theta = y$ and $\cos\theta = x$, we see that $\tan\theta = \frac{\sin\theta}{\cos\theta}$.

So, for right triangles inscribed in the circle of radius 1 centered at (0,0) the sine values are equal to the value of the y coordinate of the point P and the cosine values are equal to the value of the x coordinate of the point P.

When θ equals zero, the y coordinate is zero and as θ increases (the ray OP rotates counterclockwise) the sine value (y coordinate) increases until it reaches a maximum value of 1 when θ = 90°. As θ increases further, the value of the y coordinate starts to decrease from 1 until it becomes zero when θ = 180°. As θ increases from 180° the y coordinate decreases from zero until it reaches its smallest value of -1 when θ = 270°. As θ increases from 270° the y coordinate increases from -1 until it equals zero again when θ equals 360°.

When θ equals zero, the cosine, which is equal to the x coordinate, is 1 and as θ increases the value of the x coordinate decreases until it is equal to zero when θ equals 90°. As θ increases from 90° the x coordinate decreases from zero until it reaches its least value of -1 when θ = 180°. As θ increases from 180° its x-coordinate starts to increase until it becomes zero when θ = 270°. And as θ increases from 270° the x coordinate increases from zero until it becomes 1 again when θ = 360°.

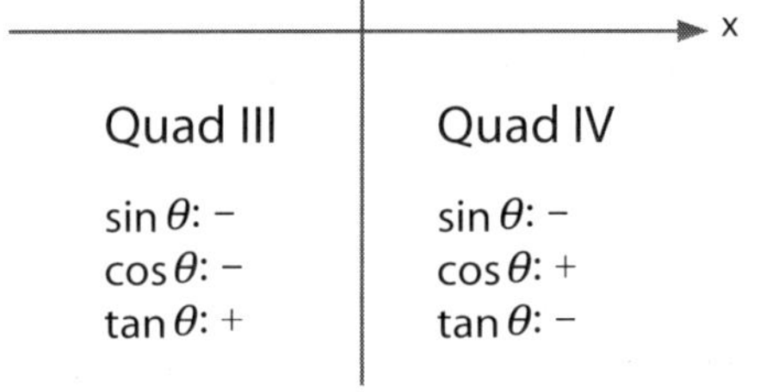

Section 5.5

Reference Angles

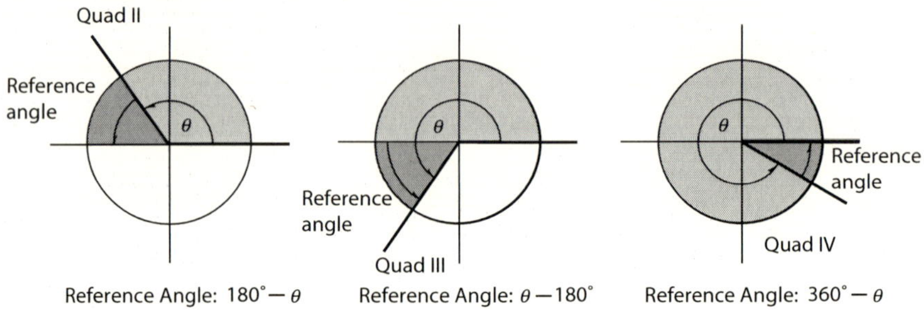

If the angle, θ, that OP makes with the positive x axis is greater than 90° and less than 180° then it lies in the second quadrant. If θ is greater than 180° and less than 270°, then it lies in the third quadrant and if θ is greater than 270° and less than 360°, then θ lies in the fourth quadrant. For angles in the second, third and fourth quadrants we calculate their sine, cosine and tangent using the angle's corresponding reference angle.
The reference angle for an angle in the second quadrant is defined to be equal to $180° - \theta$. The reference angles for angles in the third and fourth quadrants are $\theta - 180°$ and $360° - \theta$ respectively.

For example, an angle θ equal to 120° would lie in the second quadrant and it would therefore have a reference angle of 180° -120° equal to 60°. The magnitude of the sine of 120° would therefore equal the magnitude of the sine of its reference angle, 60°. And since $\sin\theta = y$ and y is positive for angles in the second quadrant we have

$$\sin 120° = \sin(180° - 120°) = \sin 60° = \frac{+\sqrt{3}}{2}.$$

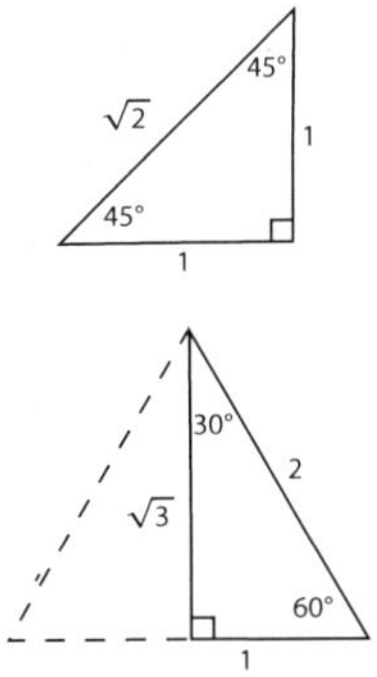

COMMON FIRST QUADRANT ANGLES

Degrees	0	30°	45°	60°	90°
Radians	0	$\frac{\pi}{6}$	$\frac{\pi}{4}$	$\frac{\pi}{3}$	$\frac{\pi}{2}$
sin θ	0	$\frac{1}{2}$	$\frac{\sqrt{2}}{2}$	$\frac{\sqrt{3}}{2}$	1
cos θ	1	$\frac{\sqrt{3}}{2}$	$\frac{\sqrt{2}}{2}$	$\frac{1}{2}$	0
tan θ	0	$\frac{1}{\sqrt{3}}$	1	$\sqrt{3}$	undef.

If we rotate the ray OP one complete revolution plus another θ degrees, then we call θ the principal angle of the total angle of rotation. For example, if we rotate OP 390°, then the principal angle of rotation for the 390° is 30°.

So to calculate the sine of 390° we first find its principal angle, which is 30° and therefore sine 390° = sine 30° = 1/2.

If we rotate OP, say, 495°, then the principal angle is 135° which lies in the second quadrant and so we must now calculate the reference angle for the 135° principal angle.

As we saw earlier, the reference angle for an angle in the second quadrant is $180° - \theta$ or in this case, 180°-135°.

So to calculate the cosine of 495° we calculate the cosine of 180° -135° = $\cos 45° = -\frac{1}{\sqrt{2}}$. The cosine is negative because cosine is negative for angles in the second quadrant.

We use the concept of the principal angle anytime we rotate OP (clockwise or counterclockwise) any multiple of one revolution plus an additional θ degrees. If we were to rotate OP two revolutions plus another 30° in the clockwise direction (for a total of -750°) the principal angle would be -30° which would lie in the fourth quadrant and $\sin(-750°)$ would equal $\sin(-30°) = -\frac{1}{2}$ since sine values are negative for angles in the fourth quadrant.

Chapter 6

Inverse Trigonometric Functions

Function: A function from set A to set B is a rule that assigns to each element, *x*, in set A a unique element, *y*, in set B.

We write $y = f(x)$ to indicate that the element y is the value assigned by the function f to the element x.

The set A is called the domain of the function f and the set of all values y is called the range of the function

Section 6.1

ArcSin Function

For the sine function, the domain (set A) is just the set of all real numbers $(-\infty < \theta < \infty)$ because we can input any degree angle no matter how large or small. However, we just saw that the output of the sine function has a maximum value of +1 and a minimum value of -1 and takes on every value in between -1 and +1.

Therefore the range of the sine function is the interval $-1 \le \sin\theta \le 1$ or $[-1,1]$.

What if we ask the question "What is the angle whose sine is ½ ?". We know that $\sin 30° = \frac{1}{2}$ so one answer would be the angle $\theta = 30°$. But since the sine of 150° is also ½, another answer could be $\theta = 150°$.

If we define ArcSin x (or $\sin^{-1} x$) to be the function equivalent of the question "What is the angle whose sine is x ?", we see there is a problem because ArcSin [½] has multiple answers but the definition of a function requires that ArcSin can only assign one unique value to x. We accomplish this by restricting the range of the ArcSin function to be the interval [−90°, 90°] or $-90° \le \theta \le 90°$.

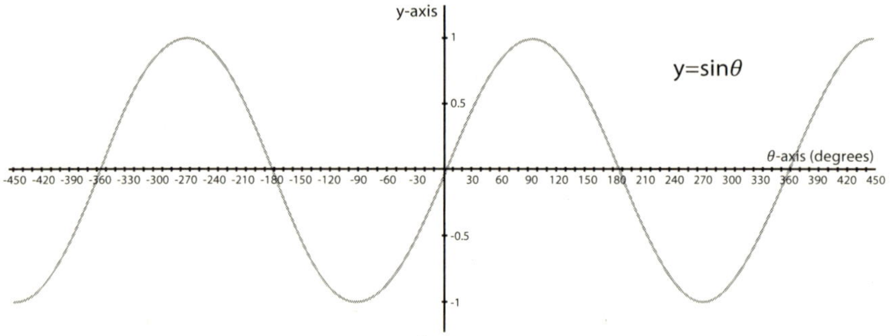

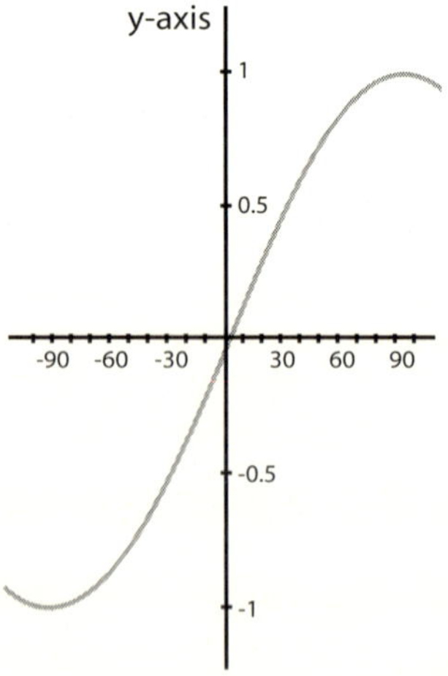

If we look at the sine function on the restricted domain of $-90° \le \theta \le 90°$, the output, or range, of the sine function still ranges from -1 (when $\theta = -90°$) to 1 (when $\theta = 90°$) and passes through 0 (when $\theta = 0°$) but the sine function on this restricted domain is one-to-one meaning that a specific value, θ, between -90° and 90° is associated with only *one* value of $\sin\theta$ in the interval $[-1,1]$.

I.e., if $\theta_1 \neq \theta_2$, then $\sin\theta_1 \neq \sin\theta_2$. If we were to extend the domain to, say, [−90°, 91°], then the sine function would no longer be one-to-one because $\sin 89° = \sin 91°$.

Now, when we ask the question "What is the angle whose sine is ½ ?". There's only one value of θ in the restricted range of the ArcSin function [−90°, 90°] where $\sin\theta = \frac{1}{2}$ and that is $\theta = 30°$.

So the unique answer to ArcSin[1/2] is $\theta = 30°$. The ArcSin function takes a value, x, between -1 and 1 and asks "What value of θ between -90° and 90° has a sine value equal to x ?". The domain of the ArcSin function is defined to be $-1 \le x \le 1$ and its range is $-90° \le \theta \le 90°$. If we did not restrict the range of the ArcSin function to $-90° \le \theta \le 90°$, but rather to $-90° \le \theta \le 91°$, for example, then $\sin 89° = x$ and also $\sin 91° = x$. That is, ArcSin[x] could be both $\theta = 89°$ and $\theta = 91°$ which contradicts the definition of a function which says that for any value in its domain a function can output only one unique value.

Defined with the restricted range of $-90° \leq \theta \leq 90°$ ArcSin is the inverse of the sine function.

(See also the discussion of inverse functions on pages 106.)

If $\sin\theta = x$, we can find out what value θ is by taking the ArcSin of both sides of this equation to get

$$\text{ArcSin}[\sin\theta] = \text{ArcSin}[x].$$

ArcSin simply "undoes" the Sin function to return to the value θ so we have

$\theta = \text{ArcSin}[x]$.

For the specific example $\sin\theta = \frac{1}{2}$ we would get

$$\text{ArcSin}[\sin\theta] = \text{ArcSin}[1/2]$$

$$\theta = \text{ArcSin}[1/2].$$

And since $\text{ArcSin}\left[\frac{1}{2}\right] = 30°$, $\theta = 30°$.

Section 6.2

ArcCos Function

The ArcCos and ArcTan functions are defined similarly.

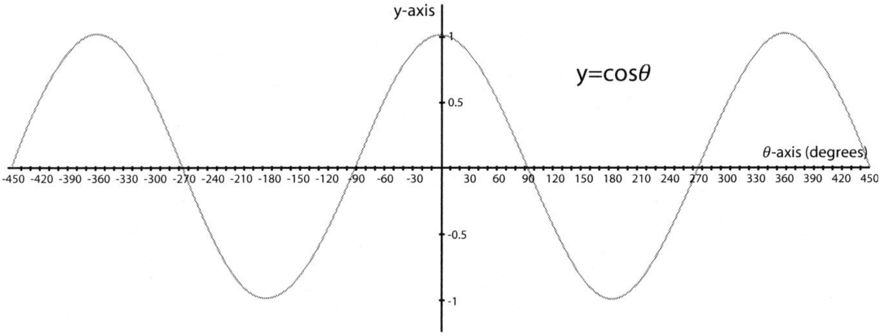

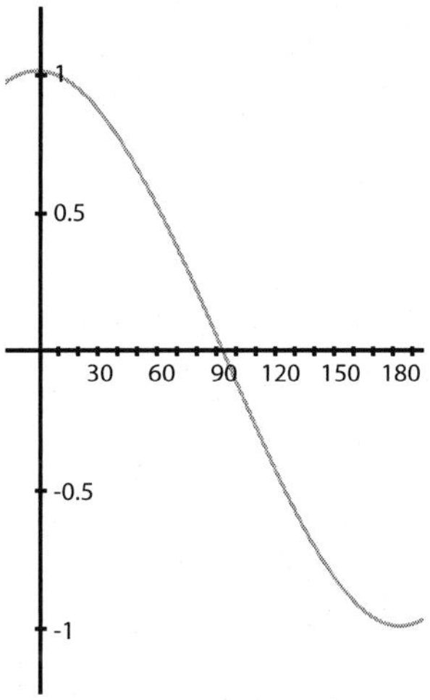

If we restrict the domain of the cosine function to the interval $0° \le \theta \le 180°$, then it is one-to-one on this interval because for $\theta_1 \neq \theta_2$, $\cos\theta_1 \neq \cos\theta_2$.

As you can see from the graph above the range of $\cos\theta$

associated with the restricted domain of $0° \leq \theta \leq 180°$ is still $[-1,1]$ meaning that $\cos\theta$ still takes on all values between +1 and -1 as θ ranges from 0 to 180°.

We now define the ArcCos function to have $-1 \leq x \leq 1$ as its domain and its range is $0° \leq \theta \leq 180°$. Defined this way, ArcCos[x] represents the question "What is the angle between 0° and 180° whose cosine is x ?".

If $\cos\theta = x$, we can find out what value θ is by taking the ArcCos of both sides of this equation to get

$$\text{ArcCos}[\cos\theta] = \text{ArcCos}[x].$$

And since ArcCos simply undoes the cosine function to return to the value θ we have $\theta = \text{ArcCos}[x]$.

For the specific example $\cos\theta = \frac{1}{\sqrt{2}}$ we would get

$$\text{ArcCos}[\cos\theta] = \text{ArcCos}[1/\sqrt{2}]$$

$$\theta = \text{ArcCos}[1/\sqrt{2}]$$

and since the angle between 0° and 180° whose cosine is $\frac{1}{\sqrt{2}}$ is the angle $\theta = 45°$

$$\theta = \text{ArcCos}\left[\frac{1}{\sqrt{2}}\right] = 45°.$$

ArcTan Function

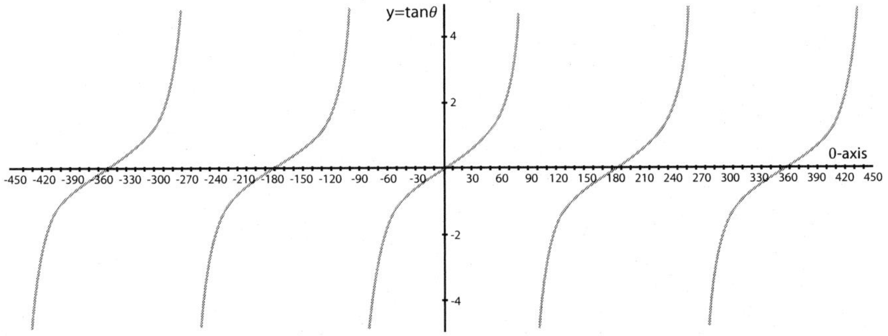

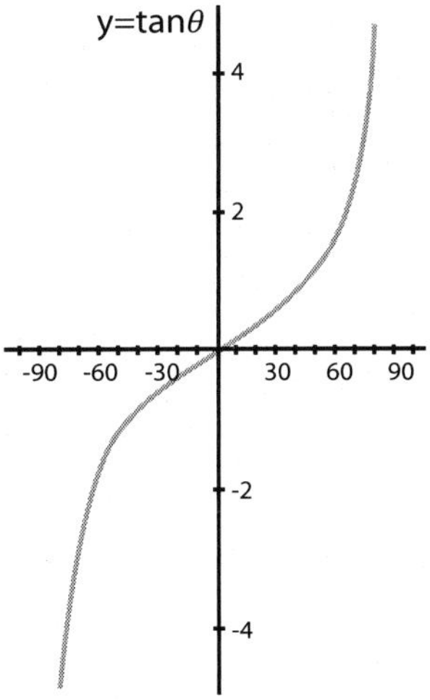

Finally, if we restrict the domain of the tangent function to the interval $-90° < \theta < 90°$, then it is one-to-one on this interval and as we can see from the graph above that its range is still $(-\infty, \infty)$.

We now define the ArcTan function to have $(-\infty < x < \infty)$ as its domain and its range is $-90° < \theta < 90°$.

Defined this way ArcTan[x] represents the question "What is the angle between −90° and 90° whose tangent is x ?"

If $\tan\theta = x$, we can find out what value θ is by taking the ArcTan of both sides of this equation to get

$$\text{ArcTan}[\tan\theta] = \text{ArcTan}[x].$$

And since ArcTan simply undoes the tangent function to return to the value θ we have $\theta = \text{ArcTan}[x]$.

For the specific example $\tan\theta = \surd 3$ we would get

$$\text{ArcTan}[\tan\theta] = \text{ArcTan}[\surd 3]$$

$$\theta = \textbf{ArcTan}[\surd \mathbf{3}]$$

and since the angle between −90° and −90° whose tangent is $\surd 3$ is the angle $\theta = 60°$

$\theta = \text{ArcTan}[\surd 3] = 60°$

Chapter 7
Pythagorean Theorem and the Law of Cosines

Section 7.1

Pythagorean Theorem

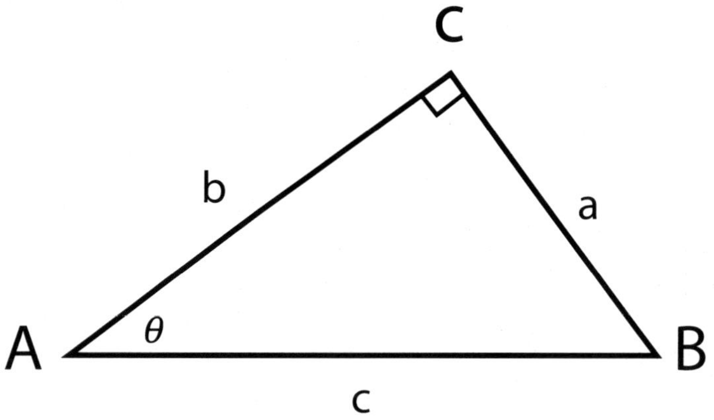

From the figure of the right-angle triangle above we see that

$\sin\theta = \frac{opp.}{hyp.} = \frac{a}{c}$ and $\cos\theta = \frac{adj.}{hyp.} = \frac{b}{c}$.

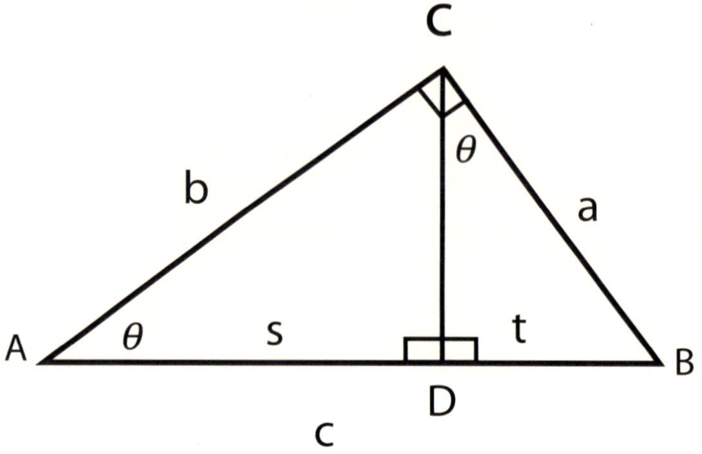

We now draw a line, within this same right triangle, from the vertex C down to point D so that the line makes a right angle with side c as shown in the figure above.

Since the interior angles of a triangle must add up to 180° and one of the angles is 90° and another is θ, we must have an angle of 180° $-90° - \theta° = 90° - \theta$ for the vertex formed by AC and CD.

And since the angle formed by AC and CB is 90°, the angle formed by DC and CB must be $90° - (90° - \theta) = \theta$.

From the left-most inscribed triangle we have

$$\cos\theta = \frac{s}{b}$$

and from the right-most triangle we have

$$\sin\theta = \frac{t}{a} .$$

So we have

$\sin\theta = \frac{a}{c}$ and $\sin\theta = \frac{t}{a}$

and

$\cos\theta = \frac{b}{c}$ and $\cos\theta = \frac{s}{b}$.

Therefore $\frac{a}{c} = \frac{t}{a}$ or $\frac{a^2}{c} = t$

and $\frac{b}{c} = \frac{s}{b}$ or $\frac{b^2}{c} = s$.

We can see from the last figure that $s + t = c$ and therefore

$$\frac{b^2}{c} + \frac{a^2}{c} = c.$$

And multiplying both sides of the equation by c we get

$$c \cdot \left(\frac{a^2}{c} + \frac{b^2}{c}\right) = c \cdot c$$

$$= a^2 + b^2 = c^2$$

which is the equation of the Pythagorean Theorem.

Law of Cosines

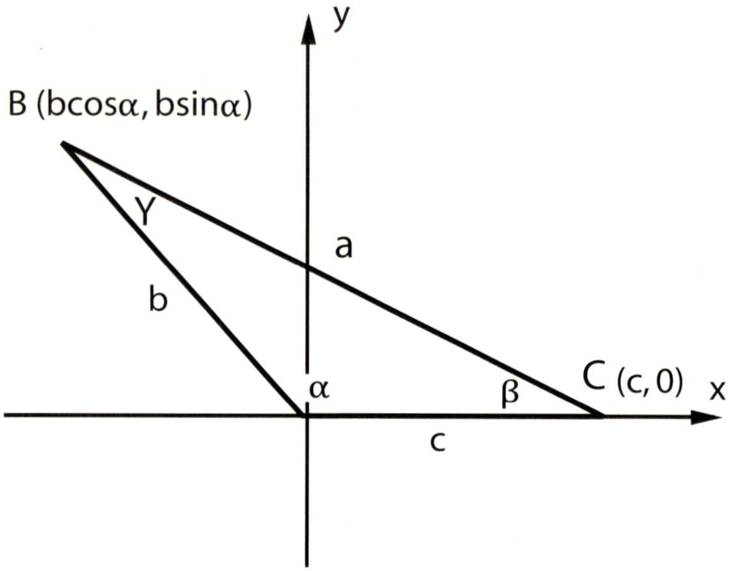

Using the formula for the distance between two points (see p. 128) we have

$$a = \sqrt{(b \cdot \cos\alpha - c)^2 + (b \cdot \sin\alpha - 0)^2}\,.$$

Squaring both sides we get

$$a^2 = (b \cdot \cos\alpha - c)^2 + (b \cdot \sin\alpha - 0)^2$$

$$= b^2 \cdot (\cos\alpha)^2 - 2bc \cdot \cos\alpha + c^2 + b^2 \cdot (\sin\alpha)^2$$

$$= b^2 \cdot (\cos\alpha)^2 + b^2 \cdot (\sin\alpha)^2 - 2bc \cdot \cos\alpha + c^2$$

$$= b^2 \cdot [(\cos\alpha)^2 + (\sin\alpha)^2] - 2bc \cdot \cos\alpha + c^2$$

$$= b^2 - 2bc \cdot \cos\alpha + c^2$$

$= b^2 + c^2 - 2bc \cdot \cos\alpha$ which is the law of cosines.

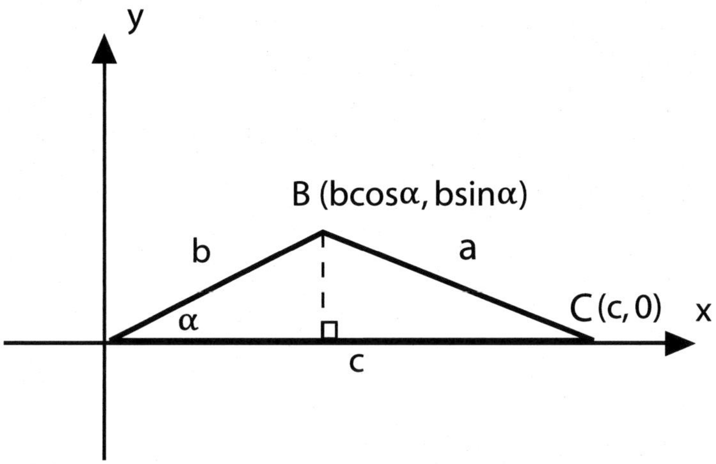

$$a = \sqrt{(b \cdot \cos\alpha - c)^2 + (b \cdot \sin\alpha - 0)^2}$$

Squaring both sides we get

$$a^2 = (b \cdot \cos\alpha - c)^2 + (b \cdot \sin\alpha - 0)^2$$

$$= b^2 \cdot (\cos\alpha)^2 - 2bc \cdot \cos\alpha + c^2 + b^2 \cdot (\sin\alpha)^2$$

$$= b^2 \cdot (\cos\alpha)^2 + b^2 \cdot (\sin\alpha)^2 - 2bc \cdot \cos\alpha + c^2$$

$$= b^2 \cdot [(\cos\alpha)^2 + (\sin\alpha)^2] - 2bc \cdot \cos\alpha + c^2$$

$$= b^2 - 2bc \cdot \cos\alpha + c^2$$

$$= b^2 + c^2 - 2bc \cdot \cos\alpha .$$

Chapter 8

Exponential and Logarithmic Functions

Section 8.1

Integer Exponents

We define 2^3 to mean 2 • 2 • 2. That is, 2 raised to the power 3 means 2 multiplied by 2 and then multiplied by 2 again. In general, if *a* is any real number and *n* is any integer greater than or equal to 1, then a^n is defined by $a^n = a \cdot a \cdot a \cdots a$ where there are a total of *n* factors of *a* on the right side of the equation. We also define a^0 (for *a* not equal to 0) to be equal to 1, and if *a* is not equal to 0 and *n* is any integer, then a^{-n} is defined to mean $\frac{1}{a^n}$. Note that a^n is not defined when $n \leq 0$ and $a = 0$.

We see that when $a \neq 0$ a^n is defined for any integer *n* but when $a = 0$, n must be greater than or equal to 1.

In the expression a^n, *a* is called the base and *n* is the exponent or power.

Section 8.2

Laws of Integer Exponents

Let *a* and *b* be real numbers and let *m* and *n* be integers. Then the following equations are true for all values of *a* and *b* where both sides of the equation are defined.

i) $a^m \cdot a^n = a^{m+n}$

ii) $(a^m)^n = a^{m \cdot n}$

iii) $(a \cdot b)^n = a^n \cdot b^n$

iv) $\left(\frac{a}{b}\right)^n = \frac{a^n}{b^n}$

v) $a^{-n} = \frac{1}{a^n}$

vi) $\frac{a^m}{a^n} = a^{m-n}$

Try a few examples using various values of *a*, *b*, *m* and *n* for each of the properties to convince yourself that they are true.

For example, $a^3 \cdot a^4 = (a \cdot a \cdot a) \cdot (a \cdot a \cdot a \cdot a) = a^7 = a^{3+4}$

and

$$a^0 \cdot a^{-3} = 1 \cdot \frac{1}{(a \cdot a \cdot a)} = \frac{1}{(a^3)} = a^{-3} = a^{0+(-3)}$$

are both consistent with property (i)

and

ii) $(a^m)^n = a^{m \cdot n}$

$$(a^{-3})^{-4} = \frac{1}{(a^{-3})^4} = \frac{1}{\left(\frac{1}{a^3}\right)^4}$$

$$= \frac{1}{\left(\frac{1}{a \cdot a \cdot a}\right)^4}$$

$$= \frac{1}{\left(\frac{1}{a \cdot a \cdot a}\right) \cdot \left(\frac{1}{a \cdot a \cdot a}\right) \cdot \left(\frac{1}{a \cdot a \cdot a}\right) \cdot \left(\frac{1}{a \cdot a \cdot a}\right)}$$

$$= \frac{1}{\frac{1}{a \cdot a \cdot a \cdot a \cdot a \cdot a \cdot a \cdot a \cdot a \cdot a \cdot a \cdot a}}$$

$$= a \cdot a \cdot a \cdot a \cdot a \cdot a \cdot a \cdot a \cdot a \cdot a \cdot a \cdot a$$

$$= a^{12} = a^{(-3) \cdot (-4)}$$

which is consistent with property (ii).

iii) $(a \cdot b)^n = a^n \cdot b^n$

$$(a \cdot b)^2 = (a \cdot b) \cdot (a \cdot b) = a \cdot b \cdot a \cdot b$$

and after applying the commutative property of multiplication of real numbers we get

$= a \cdot a \cdot b \cdot b$

$= a^2 \cdot b^2$

which is consistent with property (iii).

iv) $\left(\frac{a}{b}\right)^n = \frac{a^n}{b^n}$

Property (iv) should be clear from the way multiplication of rational numbers is defined by (1.8) on page 17.

v) $a^{-n} = \frac{1}{a^n}$

Property (v) is true by definition. For example, by definition a^{-3} means $\frac{1}{a^3}$.

vi) $\frac{a^m}{a^n} = a^{m-n}$

$$\frac{a^3}{a^{-2}} = \frac{(a \cdot a \cdot a)}{\frac{1}{(a \cdot a)}} = (a \cdot a \cdot a) \cdot \frac{(a \cdot a)}{1} = a^5 = a^{(3-[-2])}$$

which is consistent with property (vi).

Rational Exponents

We define the *n*th root of *a*, or $a^{\frac{1}{n}}$, to be the number that when raised to the power *n* is equal to the number *a*.

I.e., $\left(a^{\frac{1}{n}}\right)^n = a^{\frac{1}{n}} \cdot a^{\frac{1}{n}} \cdots a^{\frac{1}{n}} = a$ where there are *n* factors of $a^{\frac{1}{n}}$.

Since we have seen that any real number raised to an even power is nonnegative it makes no sense to ask for the *n*th root of a negative number when *n* is even. That is, when *n* is an even number $a^{\frac{1}{n}}$ is defined only for $a \geq 0$. On the other hand $a^{\frac{1}{n}}$ is defined for any real number, *a*, when *n* is odd.

For example, $(-8)^{\frac{1}{3}} = -2$ because $(-2)^3 = -8$.

We now define $a^{\frac{m}{n}}$ to be the *n*th root of *a* raised to the power *m* or $a^{\frac{m}{n}} = \left(a^{\frac{1}{n}}\right)^m$.

Here it is assumed that $\frac{m}{n}$ is in lowest terms, meaning that *m* and *n* have no common factors other than 1, and that *n* > 0 so that the sign of $\frac{m}{n}$ is determined by the sign of *m*.

Example: $8^{\frac{-4}{3}} = \left(8^{\frac{1}{3}}\right)^{-4} = 2^{-4} = \frac{1}{2^4} = \frac{1}{2 \cdot 2 \cdot 2 \cdot 2} = \frac{1}{16}$.

Notice, though, that if we first raise 8 to the power of -4, we get $8^{-4} = \frac{1}{8^4} = \frac{1}{4096}$ and then if we take the cube (third) root of this new number, $\left(\frac{1}{4096}\right)^{\frac{1}{3}}$, we get $\frac{1}{16}$ because $\frac{1}{16} \cdot \frac{1}{16} \cdot \frac{1}{16} = \frac{1}{4096}$.

So even though we defined $8^{\frac{-4}{3}}$ to be the cube root of 8 raised to the power -4 and that result is $\frac{1}{16}$, we arrived at that same number by first raising 8 to the power -4 and then taking the cube root.

$$a^{\frac{m}{n}} = \left(a^{\frac{1}{n}}\right)^m \qquad \text{(By Definition)}$$

Raising both sides of this equation to the power *n* we get

$$(a^{\frac{m}{n}})^n = \left[\left(a^{\frac{1}{n}}\right)^m\right]^n.$$

$= \left(a^{\frac{1}{n}}\right)^{m \cdot n}$ (By Property (ii) of the laws of integer exponents)

$= \left(a^{\frac{1}{n}}\right)^{n \cdot m}$ (Commutative property of real numbers)

$= \left[\left(a^{\frac{1}{n}}\right)^{n}\right]^{m}$ (By property (ii) of the law of integer exponents in reverse)

$= a^{m}$ (By definition of $a^{\frac{1}{n}}$)

So we have $(a^{\frac{m}{n}})^{n} = a^{m}$ and taking the *n*th root of both sides of this equation

we get

$a^{\frac{m}{n}} = (a^{m})^{\frac{1}{n}}$.

Section 8.4

Laws of Rational Exponents

By definition $a^{\frac{1}{3}}$ means $a^{\frac{1}{3}} \cdot a^{\frac{1}{3}} \cdot a^{\frac{1}{3}} = a$

and therefore

$$(a^{\frac{1}{3}} \cdot a^{\frac{1}{3}} \cdot a^{\frac{1}{3}}) \cdot (a^{\frac{1}{3}} \cdot a^{\frac{1}{3}} \cdot a^{\frac{1}{3}}) = a \cdot a$$

or

$$a^{\frac{1}{3}} \cdot a^{\frac{1}{3}} \cdot a^{\frac{1}{3}} \cdot a^{\frac{1}{3}} \cdot a^{\frac{1}{3}} \cdot a^{\frac{1}{3}} = a^2 \;;$$

$$(a^{\frac{1}{3}})^6 = a^2 \;.$$

In other words $a^{\frac{1}{3}}$ is also the number that when raised to the power of 6 equals a^2.

Therefore $a^{\frac{1}{3}} = a^{\frac{2}{6}}$.

Let $b = a^{\frac{2}{3}} = (a^2)^{\frac{1}{3}}$.

Then, since $(a^2)^{\frac{1}{3}}$ is the number that when raised to the power 3 is equal to a^2

$$b \cdot b \cdot b = a^2$$

and therefore

$(b \cdot b \cdot b) \cdot (b \cdot b \cdot b) = a^2 \cdot a^2 = a^4$

or

$b^6 = a^4$.

That is, $b = a^{\frac{2}{3}}$ is also the number that when raised to the power 6 is equal to a^4. Therefore $b = a^{\frac{2}{3}} = a^{\frac{4}{6}}$.

Let $b = a^{\frac{m}{n}} = (a^m)^{\frac{1}{n}}$.

Then, since $(a^m)^{\frac{1}{n}}$ is the number that when raised to the power *n* is equal to a^m

$b \cdot b \cdots b = a^m$ where there are *n* factors of *b* on the left side of the equation

and therefore

$(b \cdot b \cdots b) \cdot (b \cdot b \cdots b) \cdots (b \cdot b \cdots b) = a^m \cdot a^m \cdots a^m = a^{m \cdot c}$ where there are c factors of $(b \cdot b \cdots b)$ and c factors of a^m.

There are *nc* factors of *b* on the left side of the equation and *mc* factors of *a* on the right therefore $b^{n \cdot c} = a^{m \cdot c}$ and taking the $n \cdot c$ th root of both sides of this equation we get $b = a^{\frac{m \cdot c}{n \cdot c}}$.

That is, $b = a^{\frac{m}{n}}$ is also the number that when raised to the power $n \cdot c$ is equal to $a^{m \cdot c}$. Therefore $b = a^{\frac{m}{n}} = a^{\frac{m \cdot c}{n \cdot c}}$.

Let *a* and *b* be real numbers and let *r* and *s* be rational numbers. Then the following equations are true for all values of *a* and *b* where both sides of the equation are defined.

i) $a^r \cdot a^s = a^{r+s}$

ii) $(a^r)^s = a^{r \cdot s}$

iii) $(a \cdot b)^r = a^r \cdot b^r$

iv) $\left(\frac{a}{b}\right)^r = \frac{a^r}{b^r}$

v) $a^{-r} = \frac{1}{a^r}$

vi) $\frac{a^r}{a^s} = a^{r-s}$

Try a few examples using various values of *a*, *b*, *r* and *s* for each of the properties to convince yourself that they are true.

For example,

$a^{\frac{1}{2}} \cdot a^{\frac{2}{3}} = a^{\frac{3}{6}} \cdot a^{\frac{4}{6}}$

and therefore

$(a^{\frac{1}{2}} \cdot a^{\frac{2}{3}})^6 = (a^{\frac{3}{6}} \cdot a^{\frac{4}{6}})^6$

$= \left(a^{\frac{3}{6}}\right)^6 \cdot \left(a^{\frac{4}{6}}\right)^6$ (By property (iii) of the laws of *integer* exponents)

$= a^3 \cdot a^4$

$= a^7$.

So we have

$$(a^{\frac{1}{2}} \cdot a^{\frac{2}{3}})^6 = a^7$$

and taking the 6th root of both sides of the equation we get

$$a^{\frac{1}{2}} \cdot a^{\frac{2}{3}} = a^{\frac{7}{6}}$$

which is consistent with property (i) $a^r \cdot a^s = a^{r+s}$.

Example 2:

$$\left(a^{\frac{2}{3}}\right)^{\frac{4}{5}}$$

Let $b = \left(a^{\frac{2}{3}}\right)^{\frac{4}{5}}$ then

$$b^5 = \left(a^{\frac{2}{3}}\right)^4$$

$$b^5 \cdot b^5 \cdot b^5 = \left(a^{\frac{2}{3}}\right)^4 \cdot \left(a^{\frac{2}{3}}\right)^4 \cdot \left(a^{\frac{2}{3}}\right)^4$$

$$b^{15} = \left(a^{\frac{2}{3}}\right)^{12}$$

$$b^{15} = \left[\left(a^{\frac{1}{3}}\right)^2\right]^{12}$$

$b^{15} = \left(a^{\frac{1}{3}}\right)^{24}$ (By property (ii) $(a^m)^n = a^{m \cdot n}$ of the laws of *integer* exponents)

$b^{15} = \left(a^{\frac{1}{3}}\right)^{3 \cdot 8}$

$b^{15} = \left[\left(a^{\frac{1}{3}}\right)^{3}\right]^{8}$ (By property (ii) of the laws of integer exponents)

$b^{15} = a^{8}$

and taking the 15th root of both sides of this equation we get

$b = a^{\frac{8}{15}}$.

And since $b = \left(a^{\frac{2}{3}}\right)^{\frac{4}{5}}$

$\left(a^{\frac{2}{3}}\right)^{\frac{4}{5}} = a^{\frac{8}{15}}$

which is consistent with property (ii) $(a^r)^s = a^{r \cdot s}$.

Example 3:

$a^{\frac{2}{3}} \cdot b^{\frac{2}{3}}$

$$\left(a^{\frac{2}{3}} \cdot b^{\frac{2}{3}}\right)^3 = \left(a^{\frac{2}{3}} \cdot b^{\frac{2}{3}}\right) \cdot \left(a^{\frac{2}{3}} \cdot b^{\frac{2}{3}}\right) \cdot \left(a^{\frac{2}{3}} \cdot b^{\frac{2}{3}}\right)$$

And after some rearranging using the commutative and associative properties of multiplication of real numbers we get

$$= (a^{\frac{2}{3}} \cdot a^{\frac{2}{3}} \cdot a^{\frac{2}{3}}) \cdot (b^{\frac{2}{3}} \cdot b^{\frac{2}{3}} \cdot b^{\frac{2}{3}})$$

$$= \left(a^{\frac{2}{3}}\right)^3 \cdot \left(b^{\frac{2}{3}}\right)^3$$

$$= a^2 \cdot b^2$$

$$= (a \cdot a) \cdot (b \cdot b)$$

$$= (a \cdot b) \cdot (a \cdot b)$$

$$= (a \cdot b)^2 \,.$$

So we have

$$\left(a^{\frac{2}{3}} \cdot b^{\frac{2}{3}}\right)^3 = (a \cdot b)^2$$

and taking the third root of both sides of the equation we get

$$a^{\frac{2}{3}} \cdot b^{\frac{2}{3}} = (a \cdot b)^{\frac{2}{3}}$$

or

$$(a \cdot b)^{\frac{2}{3}} = a^{\frac{2}{3}} \cdot b^{\frac{2}{3}}$$

which is consistent with property (iii) $(a \cdot b)^r = a^r \cdot b^r$.

Example 4:

$$\left(\frac{a}{b}\right)^{\frac{2}{3}}$$

$$\left(\frac{a}{b}\right)^{\frac{2}{3}} \cdot \left(\frac{a}{b}\right)^{\frac{2}{3}} \cdot \left(\frac{a}{b}\right)^{\frac{2}{3}} = \left(\frac{a}{b}\right)^{2}$$

$$\left[\left(\frac{a}{b}\right)^{\frac{2}{3}}\right]^{3} = \left(\frac{a}{b}\right)^{2}$$

$$\left[\left(\frac{a}{b}\right)^{\frac{2}{3}}\right]^{3} = \frac{a^2}{b^2}.$$

Now taking the third root of both sides of the equation we get

$$\left(\frac{a}{b}\right)^{\frac{2}{3}} = \left(\frac{a^2}{b^2}\right)^{\frac{1}{3}}$$

$$= \frac{(a^2)^{\frac{1}{3}}}{(b^2)^{\frac{1}{3}}}$$

$$= \frac{a^{\frac{2}{3}}}{b^{\frac{2}{3}}}$$

which is consistent with property (iv) $\left(\frac{a}{b}\right)^r = \frac{a^r}{b^r}$.

Example 5:

(v) $a^{-r} = \frac{1}{a^r}$ is true by definition. For example, $a^{-\frac{2}{3}}$ is defined to be equal to $\frac{1}{a^{\frac{2}{3}}}$.

Example 6:

$$\frac{a^{\frac{7}{12}}}{a^{\frac{1}{3}}} = \frac{a^{\frac{7}{12}}}{a^{\frac{4}{12}}}$$

Let $c = \frac{a^{\frac{7}{12}}}{a^{\frac{4}{12}}}$.

Therefore $c^{12} = \frac{a^7}{a^4} = a^3$

and taking the 12th root of both sides of the equation

$c^{12} = a^3$ we get

$$c = a^{\frac{3}{12}}.$$

Therefore $c = \frac{a^{\frac{7}{12}}}{a^{\frac{4}{12}}} = \frac{a^{\frac{7}{12}}}{a^{\frac{1}{3}}} = a^{\frac{3}{12}}$

which is consistent with property (vi) $\frac{a^r}{a^s} = a^{r-s}$.

Section 8.5

Exponential Functions

Now that we know how to work with rational number exponents we will define the function $f(x) = a^x; (a > 0)$ for all real numbers x (rational and irrational).

We do this by introducing the idea of a limit. For example, $2^{\sqrt{3}}$ is defined by $\lim_{r\to\sqrt{3}} 2^r$ where r is rational. This statement is read "The limit of 2^r as r approaches $\sqrt{3}$."

What we are doing here is approximating the irrational number $\sqrt{3}$ by a rational number r as "r approaches $\sqrt{3}$."

The table below shows some decimal approximations for $\sqrt{3}$ and the corresponding values for 2^r. Note that the last row contains $\sqrt{3}$ for r and the corresponding output from my calculator for the value of $2^{\sqrt{3}}$.

r	2^r
1.5	2.828427
1.6	3.031433
1.7	3.249010
1.71	3.271608
1.72	3.294364
1.73	3.317278
1.732	3.321880
1.73205	3.321995
1.7320507	3.321996838
$\sqrt{3}$	3.321997

Notice that as r gets closer to $\sqrt{3}$ (which has a decimal expansion to 10 decimal places of 1.7320508075) 2^r gets closer to 3.321997 which is what my calculator gives for $2^{\sqrt{3}}$.

$2^{\sqrt{3}}$ is not exactly equal to 3.321997 but they are the same up to six decimal places.

Here we have used (rational) values of r that "approach" $\sqrt{3}$ from below ($r < \sqrt{3}$) but the same pattern emerges if we use values of r that approach $\sqrt{3}$ from above ($r > \sqrt{3}$).

We know that every terminating decimal number is a rational number so if we chose to, we could approximate $\sqrt{3}$ by a decimal number with, say, one thousand numbers to the right of the decimal point. This decimal number would be equal to a rational number of the form $\frac{a}{b}$ where the denominator, *b*, is equal to 10^{1000} .

It is shown in more advanced math books that for all real numbers, *x*, $\lim_{r \to x} a^r = a^x$ and this is how we define the exponential function.

Definition of Exponential Function: $a^x = \lim_{r \to x} a^r$ (where $a > 0$ and *r* is rational).

A proof can also be found in more advanced math books that when the exponential function is defined this way, the laws of exponents that were true for rational numbers can now be extended to include all real numbers *x* and *y*.

Section 8.6

Laws of Real Exponents

Let *a* and *b* be constants greater than 0. Then the

following properties are true for all real numbers x and y.

i) $a^x \cdot a^y = a^{x+y}$

ii) $(a^x)^y = a^{x \cdot y}$

iii) $(a \cdot b)^x = a^x \cdot b^x$

iv) $\left(\frac{a}{b}\right)^x = \frac{a^x}{b^x}$

v) $a^{-x} = \frac{1}{a^x}$

vi) $\frac{a^x}{a^y} = a^{x-y}$

Below are some graphs of $y = a^x$ for $0 < a < 1$ and $a > 1$.

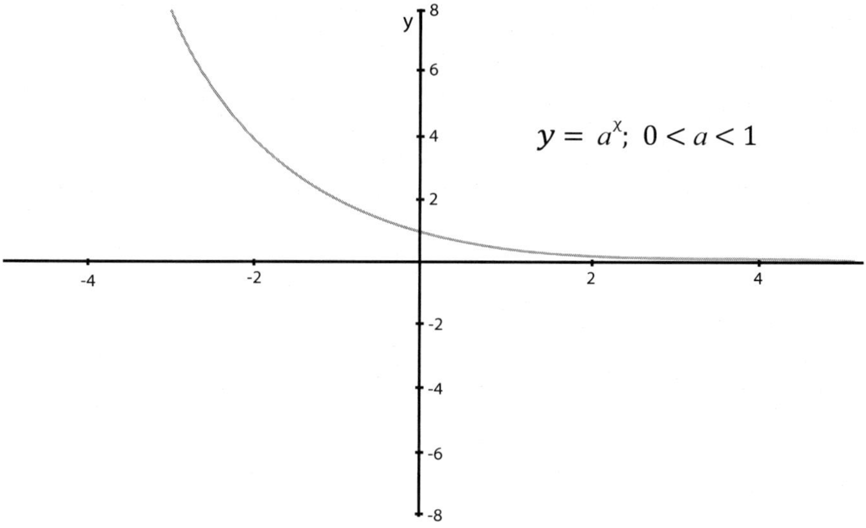

If $0 < a < 1$, then the graph of $y = a^x$ is decreasing meaning that if $x_1 < x_2$, then $a^{x_2} < a^{x_1}$.

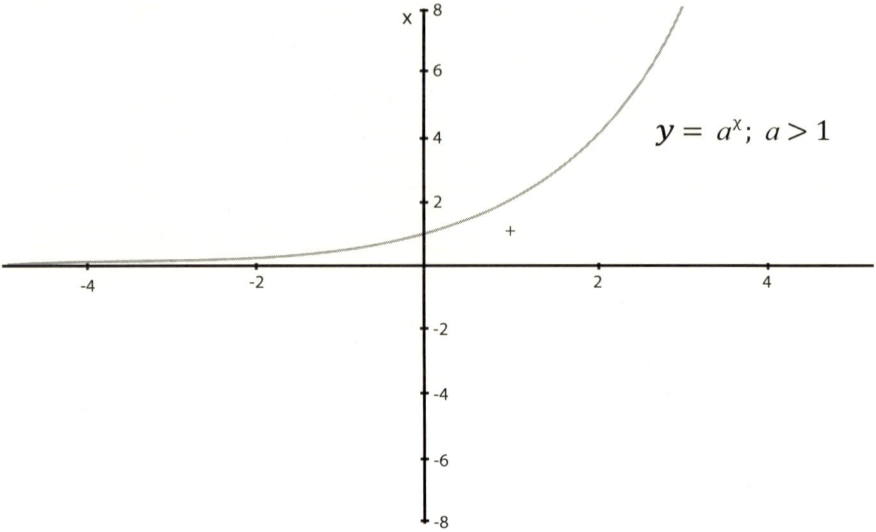

If $a > 1$, then the graph of $y = a^x$ is increasing meaning that if $x_1 < x_2$, then $a^{x_2} > a^{x_1}$.

You can also see from the graphs that (when $a \neq 1$) the exponential function has the following properties:

i) The range of $f(x) = a^x$ is the set of positive real numbers. That is, $f(x)$ is greater than 0 for all x in the domain of $f(x)$.

ii) $f(x) = a^x$ crosses the y-axis at y = 1 (this is called the y-intercept) but $f(x)$ never crosses the x-axis (no x-intercept).

iii) When $0 < a < 1$, as x tends to $+\infty$, $f(x)$ tends to 0 and as x tends to $-\infty$, $f(x)$ tends to $+\infty$. When $a > 1$, as x tends to $+\infty$, $f(x)$ also tends to $+\infty$ and as x tends to $-\infty$, $f(x)$ tends to 0.

iv) For $a > 1$: If $x_1 < x_2$, then $a^{x_1} < a^{x_2}$. That is, $f(x) = a^x$ is an increasing function.

For $0 < a < 1$: If $x_1 < x_2$, then $a^{x_2} < a^{x_1}$. That is, $f(x) = a^x$ is a decreasing function.

You can also see from the graphs that $f(x) = a^x$ is a one-to-one function. Recall that a one-to-one function is one where a value, y, in the function's range corresponds to only one value of, x, in the function's domain. In other words, $f(x_1) = f(x_2)$ only when $x_1 = x_2$ or equivalently if $x_1 \neq x_2$, then $f(x_1) \neq f(x_2)$.

Section 8.7

Logarithmic Function

Since $a^x = y$, we can ask the question "What exponent must we raise a to to get y?" and the answer is obviously x. The question can be stated in equation form as $\log_a y = x$. For example, $\log_{10} 100$ equals 2 because we must raise 10 to the power of 2 to equal 100.

$a^x = y$ and $\log_a y = x$ are really just two different forms of the same statement. If $a^x = y$, then $\log_a y = x$, and if $\log_a y = x$, then $a^x = y$. The two statements are logically equivalent and we can write this as $a^x = y$ if and only if $\log_a y = x$.

When we talked about inverse functions earlier we saw that an inverse function, f^{-1}, takes the value, y, (where $y = f(x)$) as input and returns back the value x. That is, if $f(x) = y$, then $f^{-1}(y) = x$. f^{-1} "undoes" what f has done. Note: f^{-1} is read "f-inverse".

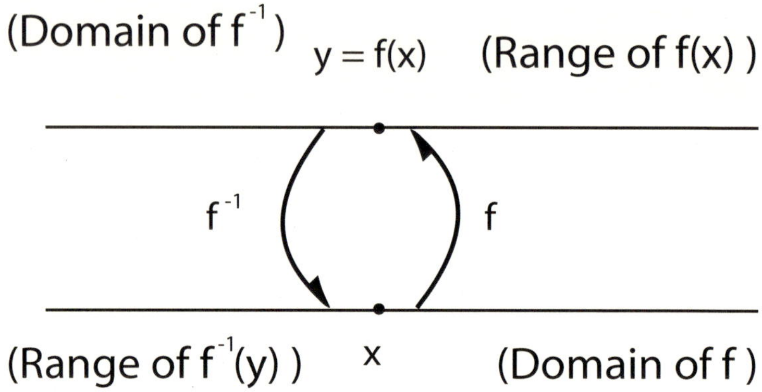

The range of $f(x)$ is equal to the domain of $f^{-1}(y)$ and the range of $f^{-1}(y)$ must lie in the domain of $f(x)$.

The domain of $f(x) = a^x$ is the set of all real numbers and the range of $f(x)$ is the set of all real numbers greater than 0. On the other hand, the domain of $f^{-1}(y) = \log_a y$ is the set of real numbers greater than 0 and the range of $f^{-1}(y)$ is the set of all real numbers.

The range of $f(x) = a^x$ corresponds to the domain of $f^{-1}(y) = \log_a y$ and the range of $f^{-1}(y) = \log_a y$ corresponds to the domain of $f(x) = a^x$.

If $a^x = c$, then $\log_a c = \log_a a^x = x$. That is, the logarithm function takes the output of the exponential function and returns back the value x which was the input for the exponential function. And if $\log_a y = d$, then $a^d = y$ and therefore $a^{\log_a y} = y$. That is, the exponential function takes the output of the logarithm function and returns back the value y which was the input for the logarithm function. This shows that the exponential and logarithmic functions are inverses of one another.

$f(x) = a^x = y$ and $f^{-1}(y) = \log_a y$ are inverses of one another because when we take a value, x, in the domain of a^x we get an output of $y = a^x$ in the range of $f(x) = a^x$. Now $f^{-1}(y) = \log_a y$ takes this value of y which lies in its domain and outputs back to the value x which lies in its range.

Section 8.8

Laws of Logarithms

Let a be a constant greater than 0 and not equal to 1. Then the following properties are true for all numbers $x > 0$ and $y > 0$.

i) $\log_a(x \cdot y) = \log_a x + \log_a y$

ii) $\log_a\left(\frac{1}{x}\right) = -\log_a x$

iii) $\log_a\left(\frac{x}{y}\right) = \log_a x - \log_a y$

iv) $\log_a(x^c) = c \cdot \log_a x$

v) $\log_a 1 = 0$

vi) $\log_a a = 1$

Proof:

i) Let $r = \log_a x$ and let $s = \log_a y$. Then we know that $a^r = x$ and that $a^s = y$ and therefore $x \cdot y = a^r \cdot a^s$. By the law of exponents $a^r \cdot a^s = a^{r+s}$ so we have $x \cdot y = a^{r+s}$. Taking the logarithm of both sides of the equation $x \cdot y = a^{r+s}$ we get

$$\log_a(x \cdot y) = \log_a(a^{r+s}) = r + s\ .$$

And since $r = \log_a x$ and $s = \log_a y$

$\log_a(x \cdot y) = \log_a x + \log_a y$ which proves property (i).

We will prove property (iii) before property (ii).

iii) Let $r = \log_a x$ and let $s = \log_a y$. Therefore $a^r = x$ and $a^s = y$.

$\frac{x}{y} = \frac{a^r}{a^s} = a^{r-s}$.   (By the law of exponents)

Taking the logarithm of both sides of the equation $\frac{x}{y} = a^{r-s}$ we get

$$\log_a\left(\frac{x}{y}\right) = \log_a(a^{r-s}) = r - s\,.$$

And since $r = \log_a x$ and $s = \log_a y$

$\log_a\left(\frac{x}{y}\right) = \log_a x - \log_a y$ which proves property (iii).

ii) We will now use property (iii) to prove property (ii).

$\log_a\left(\frac{1}{x}\right) = \log_a 1 - \log_a x$ and since $\log_a 1 = 0$ we have

$\log_a\left(\frac{1}{x}\right) = 0 - \log_a x = -\log_a x$ which proves property (ii).

iv) $\log_a(x^c) = c \cdot \log_a x$

Let $r = \log_a x$. Therefore $a^r = x$ or $x = a^r$. Raising both sides of this equation to the power c we get

$x^c = (a^r)^c = a^{r \cdot c}$.   (By the law of exponents)

Now, taking the logarithm of both sides of $x^c = a^{r \cdot c}$ we get

$$\log_a x^c = \log_a(a^{r \cdot c}) = r \cdot c = c \cdot r\,.$$

That is,

$\log_a x^c = c \cdot r$ and since $r = \log_a x$ we have

$\log_a x^c = c \cdot \log_a x$ which proves property (iv).

v) $\log_a 1 = 0$

$\log_a 1 = 0$ since a^0 is defined to be equal to 1.

vi) $\log_a a = 1$

$\log_a a = 1$ since $a^1 = a$.

We can extend these laws of logarithms to three or more variables. For example,

$\log_a x \cdot y \cdot z = \log_a x \cdot y + \log_a z$ by property (i)

$= \log_a x + \log_a y + \log_a z$ by property (i) again.

and

$\log_a \frac{x \cdot y}{z} = \log_a x \cdot y - \log_a z$ by property (ii)

$= \log_a x + \log_a y - \log_a z$. by property (i).

Section 8.9

Change of Base of Logarithm

Suppose that $\log_b x = r$ and we want to find $\log_a x$ where a and b are two different positive numbers not equal to 1. Since $\log_b x = r$, $b^r = x$ or $x = b^r$. Taking the logarithm to the base a of both sides of $x = b^r$ we get

$$\log_a x = \log_a b^r$$

$= r \cdot \log_a b$. (by property (iv) of logarithms)

Since $b \neq 1$, $\log_a b \neq 0$ and we can divide both sides of

$r \cdot \log_a b = \log_a x$ by $\log_a b$

to get

$$r = \frac{\log_a x}{\log_a b}.$$

Now, since $r = \log_b x$ we have

$$\log_b x = \frac{\log_a x}{\log_a b}$$ or

$$\log_a x = \log_b x \cdot \log_a b$$.

Appendix A

Proof that $\sqrt{2}$ is an irrational number

The method of proof we will use to show that $\sqrt{2}$ is irrational is called "Proof by Contradiction." We will assume that $\sqrt{2}$ is a rational number and then using a series of valid mathematical operations come to a contradiction showing that our assumption was false.

Since we assume that $\sqrt{2}$ is rational it can be written in the form

$\sqrt{2} = \frac{a}{b}$ where *a* and *b* are integers that have no common factors and $b \neq 0$.

If *a* and *b* did have common factors, we could factor them out and cancel to get new values of *a* and *b* that have no common factors.

$\sqrt{2} = \frac{a}{b}$ and squaring both sides we get

$2 = \frac{a^2}{b^2}$ multiplying both sides by b^2 we get

$2b^2 = a^2$.

Since $2b^2$ has a factor of 2 it is an even integer and therefore a^2 is also an even integer.

Since the only way to obtain an even integer by squaring is to square an even integer a must be even and we can write it as

$a = 2 \cdot c$ for some integer c.

Thus from $2b^2 = a^2$ we get

$$2b^2 = (2 \cdot c)^2$$

$2b^2 = 4 \cdot c^2$ and after dividing both sides by 2 we get

$$b^2 = 2 \cdot c^2.$$

Since $2 \cdot c^2$ is an even integer so is b^2 and therefore b is also even.

We have shown that both a and b are even integers but they can't both be even and not have a common factor as we assumed. Thus our assumption that $\sqrt{2}$ is a rational number is false and we conclude that $\sqrt{2}$ is irrational.

Appendix B

Radian Measure of Angles

Even though we have used the degree measurement of an angle throughout this book the radian unit of angular measurement is much more commonly used in math and science. Suppose that we have a circle with radius = r. If we were to take some string whose length is also equal to r and lay the string along the circumference of the circle, then the angle of the circle that the string subtends is defined to be equal to one radian.

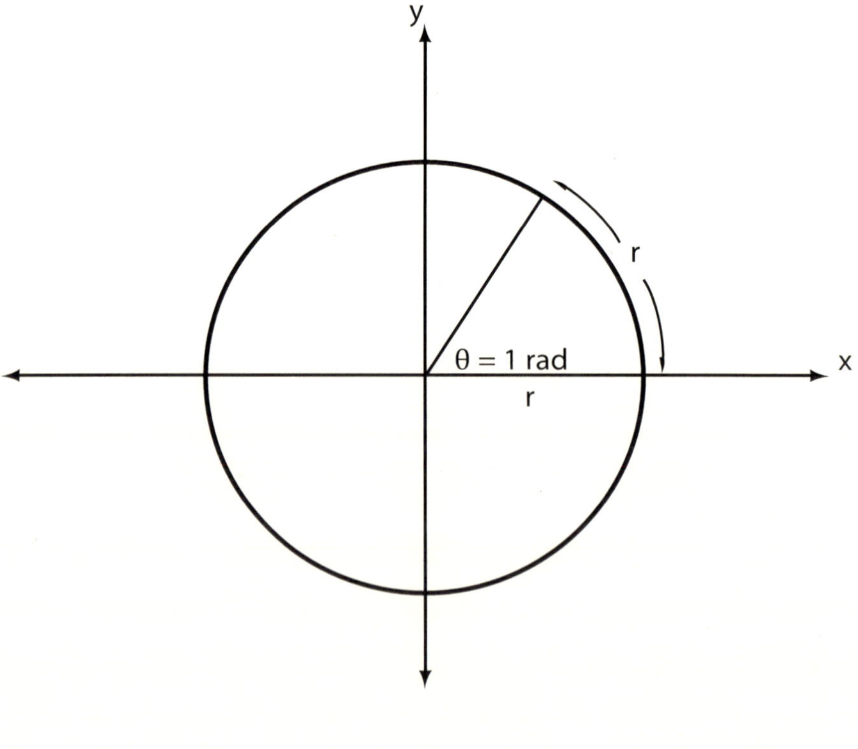

Since the entire circumference of the circle has a length equal to $2\pi \cdot r$ there are a total of 2π radians associated with one revolution of the circle. There are 2π radians per $360°$ or we can also say that there are $360°$ associated with 2π radians.

$$\frac{2\pi \text{ radians}}{360°};$$

$$\frac{360°}{2\pi \text{ radians}}.$$

$360°$ per 2π radians means that there are ~ 57.3 degrees per radian. Radians are a pure number and have no units.

Since we know that there is an arc length equal to r associated with one radian we can calculate the arc length associated with any multiple of radians. For example, if we have $r = 10$ inches and want to know what arc length is associated with an angle measure of 1.5 radians then we would have

$$\left(\frac{10 \text{ inches}}{\text{radian}}\right) \cdot (1.5 \text{ radians}) = 15 \text{ inches}.$$

If S represents arc length, then $S = (\text{radius}) \cdot (\text{angle in radians})$.

Appendix C

Rectangular Coordinate System

The real number line represents just one dimension and takes only one number (or coordinate) to specify a specific point on the line. We can also plot points in two dimensional space using the rectangular coordinate system. The rectangular coordinate system, or Cartesian plane, consists of two perpendicular real lines that intersect at the origin of the coordinate system as shown below.

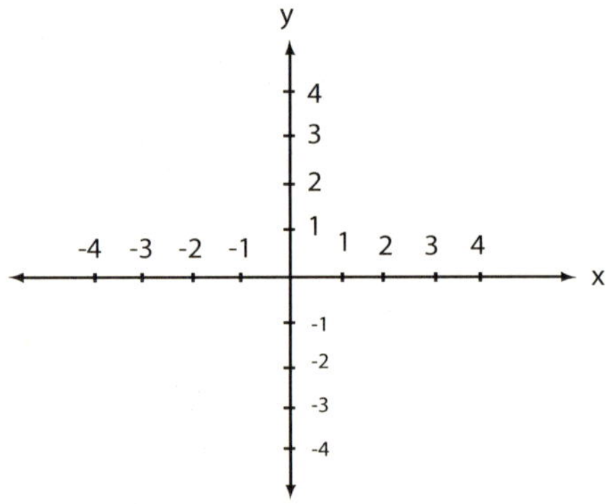

The x axis is the same as the real number line and is perpendicular to the y axis. Positive values of y are above the x axis and negative values are below the x axis. Each point in the coordinate plane (which is also called the x-y coordinate plane) is specified by an ordered pair of real numbers (x,y). For example, the point (1,2) would lie a distance of one unit to the right of the y-axis and a distance of two units above the x-axis as shown.

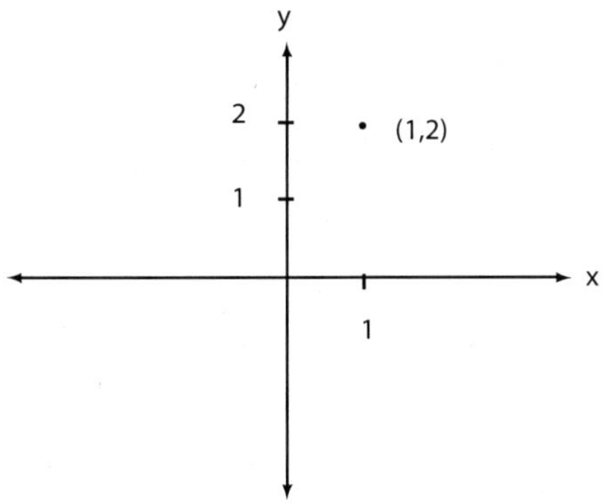

(x,y) is called an ordered pair because the order in which the numbers occur matters. For example, (1,2) is a different point from (2,1).

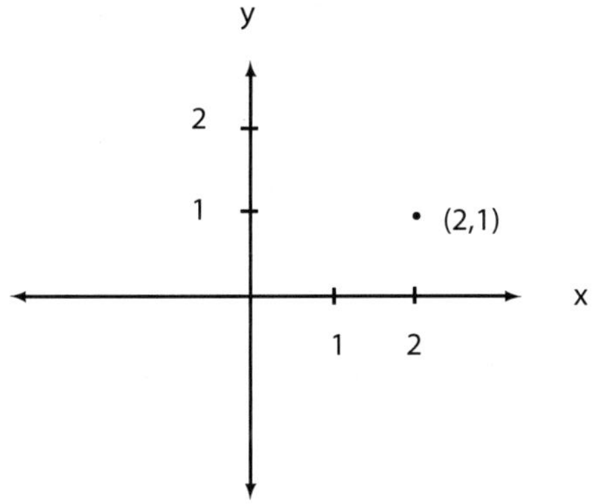

The simplest way to graph a function $y = f(x)$ is to specify the values for x and then compute the corresponding values of y and plot the coordinate pair.

As a simple example we plot $y = 2x + 1$ using this method.

x	$2x + 1$
-3	-5
-2	-3
-1	-1
0	1
1	3
2	5
3	7

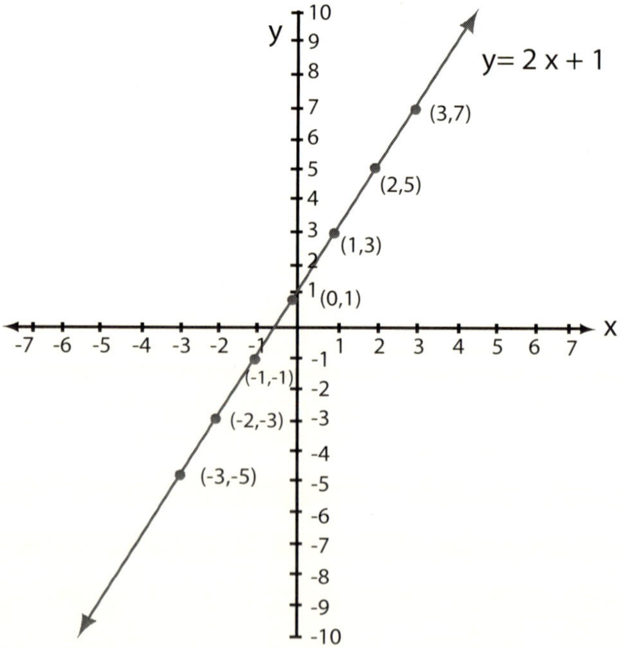

Appendix D

Vector Basics in R^2

Definition: A vector is an object that has both a magnitude and a direction.

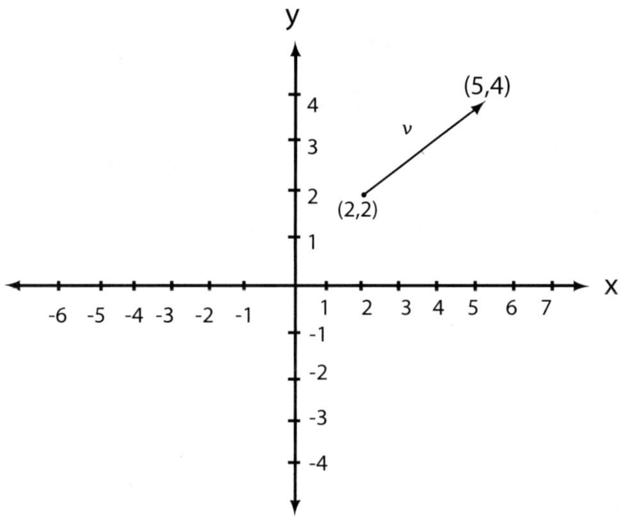

In the figure above , the vector v has its initial point at (2,2) and its terminal point at (5,4) but this is not a unique representation of this particular vector. If we translate it (slide it in the x-y plane) while maintaining its length and direction, we can position it with its initial point at any point in the plane.

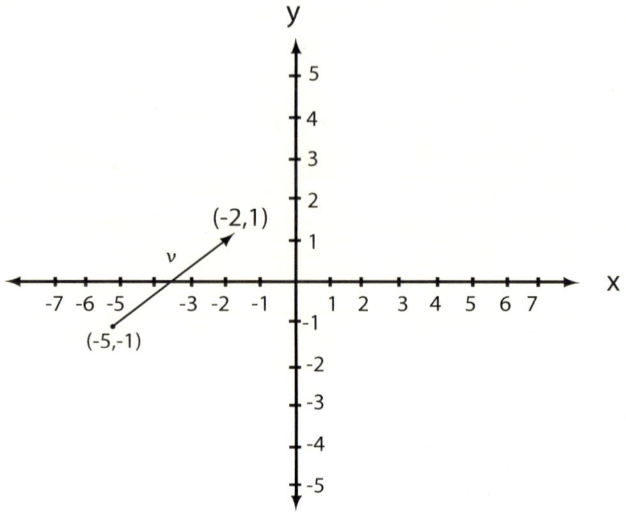

For example, we can translate it so that its initial point is at, say, (-5,-1). When this is the case its terminal point will be at (-2,1).

Because vectors are identical when they have the same magnitude and direction regardless of the position of their initial points they are sometimes called "free vectors".

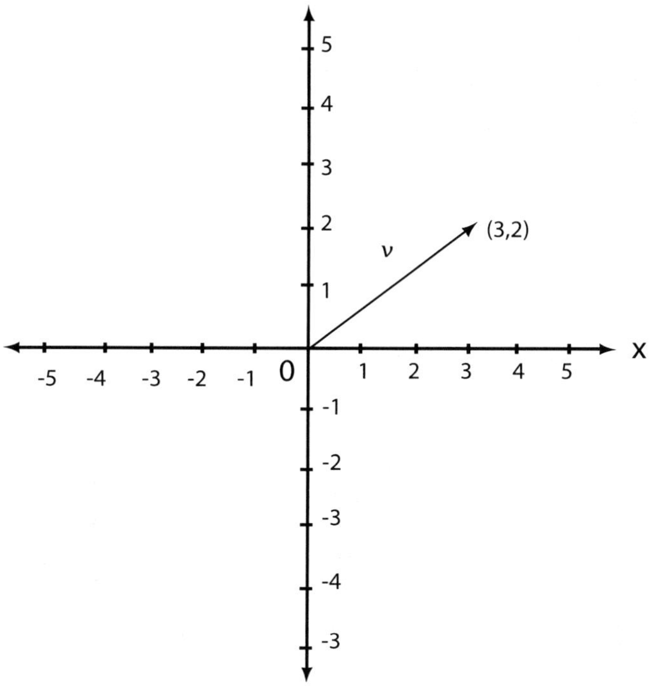

If we translate *v* so that its initial point is at the origin, then its terminal point is at (3,2).

In this case it's easy to see that the difference between the x coordinates of the vector's terminal and initial points is $3 - 0 = 3$ and the difference between the y coordinates of the vector's terminal and initial points is $2 - 0 = 2$.

Even though *v* does not have a unique geometric representation it does have a unique algebraic representation and it is the ordered pair (3,2). That is, the terminal point of *v* will always have an x coordinate that is 3 greater than the x coordinate of its initial point and the terminal point will always have a y coordinate that is 2 greater than the y coordinate of its initial point.

Definition: Addition of Vectors

If $\boldsymbol{v} = (x_1, y_1)$ and $\boldsymbol{w} = (x_2, y_2)$ are two vectors in the x-y plane, then we define

$v + w$ by

$$\boldsymbol{v} + \boldsymbol{w} = (x_1, y_1) + (x_2, y_2) = (x_1 + x_2, y_1 + y_2).$$

For example, if $v = (2,1)$ and $w = (3,5)$, then

$v + w = (2 + 3, 1 + 5) = (5, 6)$.

We say that (5, 6) is the resultant vector obtained by adding v and w.

We can represent this vector addition geometrically using the “head to tail” method.

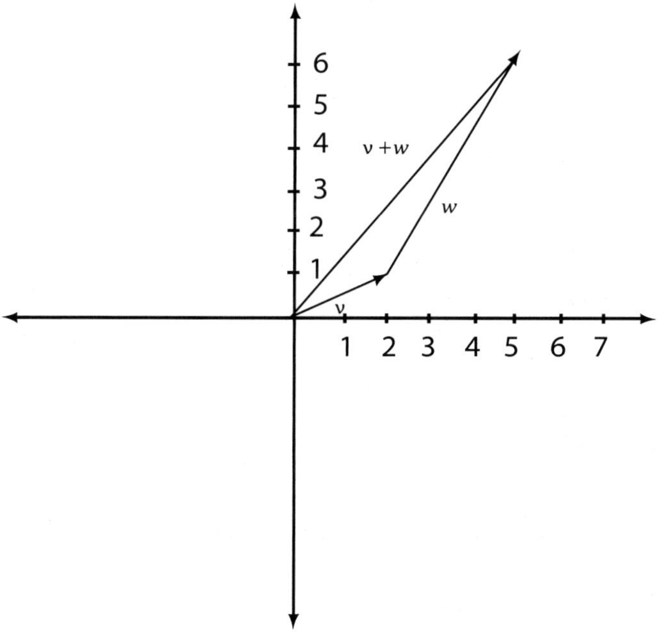

First, draw $v = (2,1)$ with its initial point (also called the tail) at the origin and its terminal point (also called its head) at (2,1). Now, placing the initial point, or tail, of $w = (3,5)$ at the head of v we see that the head of w will be at (5,6). The resultant vector is drawn from the tail of v to the head of w as shown.

When adding two vectors geometrically it is not necessary to place the tail of the first vector at the origin but you must always place the tail of the second vector at the head of the first vector. Also, the algebraic and geometric methods we used to add two vectors together is the same basic procedure used for adding three or more vectors together.

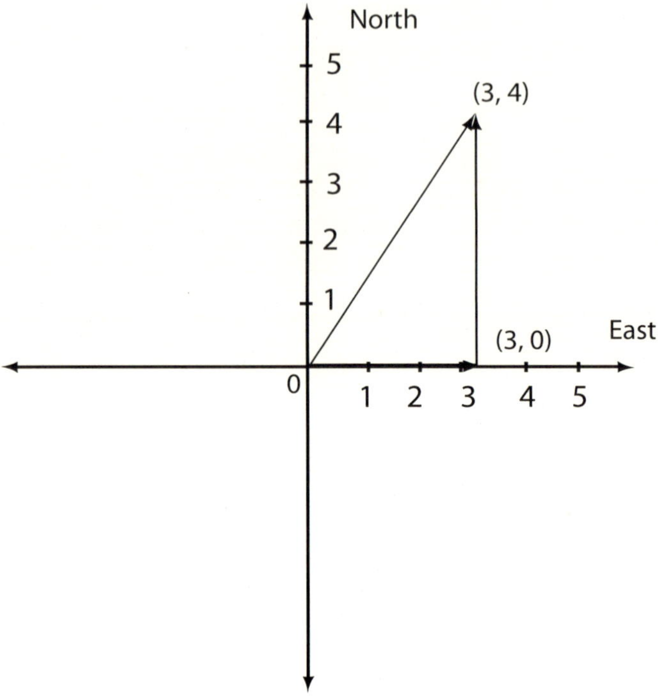

Now suppose that you were to walk due east for three feet and then stop, turn north and walk four feet due north. The first leg of your short trip would be described by the displacement vector (3, 0) and the second leg would be described by the displacement vector (0, 4). The result would be that you are now 3 feet east and four feet north of where you started. The resultant displacement of your two stage trip would be

$(3 + 0, 0 + 4) = (3,4)$ as shown in the figure above.

We can use Pythagoras' Theorem to calculate that you are $\sqrt{3^2 + 4^2} = \sqrt{25} = 5$ feet directly away from where you started. This is also how the magnitude of a vector is calculated. The magnitude of (x, y) or $|(x, y)|$ is defined by

$$|(x, y)| = \sqrt{x^2 + y^2}.$$

Since $\tan\theta = \frac{4}{3}$, $\theta = \tan^{-1}(\frac{4}{3}) = 53.1°$.

Therefore you are now 5 feet away from where you started at an angle of $53.1°$ north of east.

Had you walked directly from (0, 0) to (3, 4) the easterly or horizontal component of your trip would still be given by (3, 0) and the northerly or vertical component of your trip would still be (0, 4).

If it would have taken you, say, one second to walk the five feet directly from (0, 0) to (3, 4), then your velocity would have been 5 feet per second in the direction $53.1°$ north of east and the horizontal component of your velocity would be 3 feet per second east and the vertical component of your velocity would be 4 feet per second north.

Now suppose that you are on the bed of a flat-bed truck that is travelling due east at 3 feet per second and that you are walking due north relative to the truck at 4 feet per second. After one second of travel you will be 3 feet east of where you were one second earlier and you will also be 4 feet north of where you were one second earlier. Your resultant displacement after one second of travel is described by the vector (3, 4). As before, this means that you are now a distance of 5 feet from where you started one second earlier and at a position $53.1°$ north of east.

Since you travelled this distance of 5 feet in one second your velocity is 5 feet per second at an angle of $53.1°$ north of east. You easterly velocity of 3 feet per second combined with your northerly velocity of 4 feet per second resulted in a velocity of 5 feet per second at an angle of $53.1°$ north of east.

Had someone observed you from above walking directing from (0, 0) to (3, 4) with a constant speed of 5 ft./s and also when you were walking northward at a speed of 4 ft./s on the back of a flat-bed truck that was travelling east at 3 ft./s they would not have seen any difference in your path or speed.

Suppose that you are travelling $30°$ north of east at a speed of 10 ft./s and that your speed is increasing at a constant rate so that one second later your speed is 25 ft./s. At that instant you are travelling 15 ft./s faster than you were one second earlier. Since this increase in speed of 15 ft./s took one second to occur your acceleration is $(15\ \frac{\text{ft.}}{\text{s}})/\text{s} = 15\ \frac{\text{ft.}}{\text{s}^2}$ in the direction $30°$ north of east.

Your new speed of 25 ft./s is such that in a one second period you now travel 15 feet further than you would have in one second when your speed was only 10 ft./s. And since your direction of travel is $30°$ north of east, the easterly and northerly components of your speed are $(15 \cdot \cos 30°)$ ft./s and $(15 \cdot \sin 30°)$ ft./s greater, respectively, and since these increases in speed occurred over the same one second period the easterly and northerly components of your acceleration are

$(15 \cdot \cos 30°)\frac{\text{ft.}}{\text{s}^2}$ and $(15 \cdot \sin 30°)\frac{\text{ft.}}{\text{s}^2}$, respectively.

Experiment shows that if you apply a force on an object of 3 Newtons in the easterly direction and simultaneously apply a force of 4 Newtons in the northerly direction, then the object accelerates in the same way as if you were to apply a single force on the object of 5 Newtons in the direction of $53.1°$ north of east.

Displacement, velocity, acceleration and force all obey the vector law of addition and therefore they are all vectors.

Appendix E

Trigonometric Identities

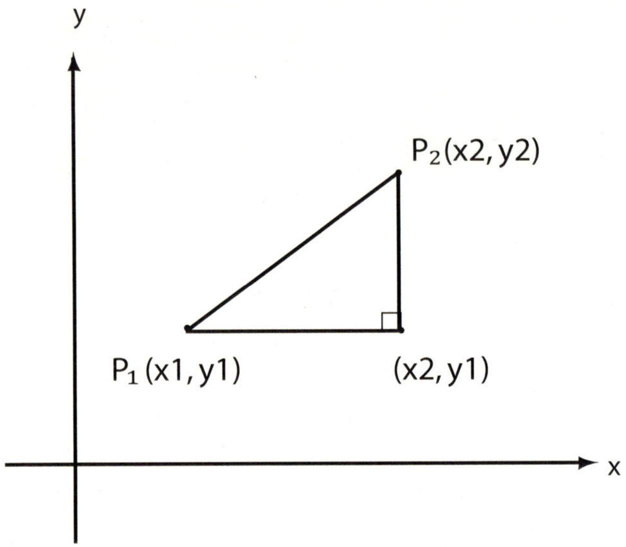

In the figure above let P_1 be the point with coordinates (x_1, y_1) and let P_2 be the point with coordinates (x_2, y_2). We can use the Pythagorean theorem to derive the distance formula for the distance between two points $P_1$and P_2.

$$(P_1P_2)^2 = (x_1 - x_2)^2 + (y_1 - y_2)^2 \text{ therefore}$$

$$P_1P_2 = \sqrt{(x_1 - x_2)^2 + (y_1 - y_2)^2} = \sqrt{(x_2 - x_1)^2 + (y_2 - y_1)^2}$$

We will now prove the basic trigonometric formula

$$\cos(\theta_1 - \theta_2) = \cos\theta_1 \cdot \cos\theta_2 + \sin\theta_1 \cdot \sin\theta_2$$

which is used to prove a number of trigonometric identities.

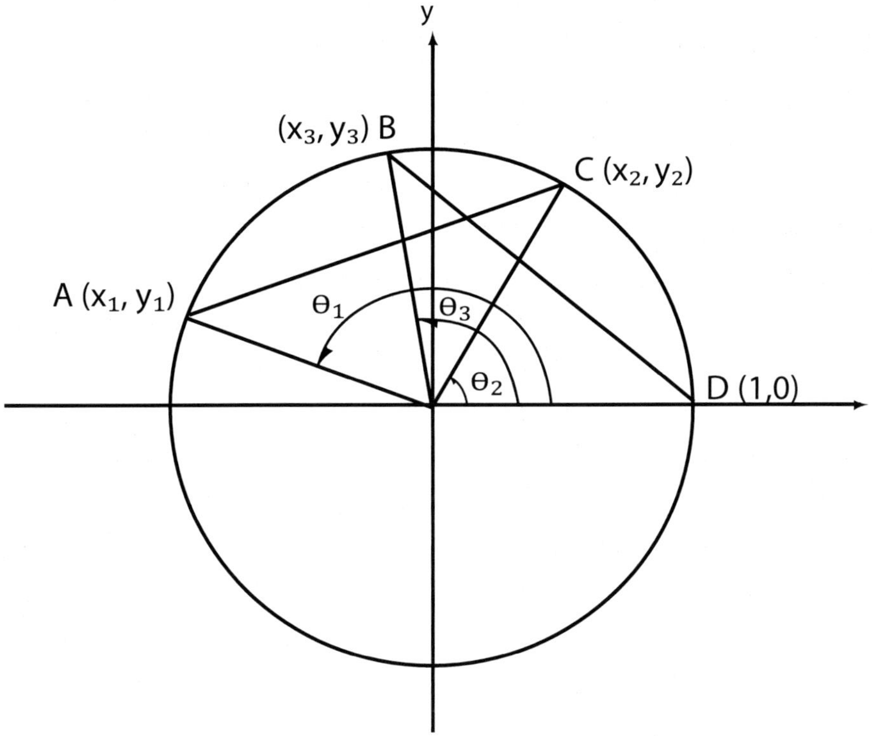

In the diagram above, angle θ_1 is the angle from the positive x axis to the point *A* which has coordinates (x_1, y_1), angle θ_3 is the angle from the positive x axis to the point *B* which has coordinates (x_3, y_3) and θ_2 is the angle from the positive x axis to the point *C* which has coordinates (x_2, y_2).

The following two conditions are also true:

i) $0 < \theta_2 < \theta_1 < 2\pi$

ii) $\theta_1 - \theta_2 = \theta_3$

We can see from the diagram that the angle between points *C* and *A* must be $\theta_1 - \theta_2$ which, according to the conditions above, is equal to θ_3. So a line drawn from point *A* to point *C* would have the same length as a line drawn from point *B* to point *D* because there is also an angle of θ_3 between point *B* and point *D*.

Therefore, we know

$x_1 = \cos\theta_1$ and $y_1 = \sin\theta_1$

$x_2 = \cos\theta_2$ and $y_2 = \sin\theta_2$

$x_3 = \cos\theta_3 = \cos(\theta_1 - \theta_2)$ and $y_3 = \sin\theta_3 = \sin(\theta_1 - \theta_2)$.

And since $(\sin\theta)^2 + (\cos\theta)^2 = 1$ for all values of θ we know that

$x_1^2 + y_1^2 = 1$, $x_2^2 + y_2^2 = 1$ and $x_3^2 + y_3^2 = 1$.

We can use the distance formula to find the distance between points *A* and *C* and also the distance between points *B* and *D* and equate them since we saw above that the two distances are equal.

The distance between *A* and *C* is given by:

$$\sqrt{(x_2 - x_1)^2 + (y_2 - y_1)^2}$$

And the distance between *B* and *D* is:

$$\sqrt{(x_3 - 1)^2 + (y_3 - 0)^2}.$$

Equating the two we get

$$\sqrt{(x_2 - x_1)^2 + (y_2 - y_1)^2} = \sqrt{(x_3 - 1)^2 + (y_3 - 0)^2}.$$

And after squaring both sides

$$(x_2 - x_1)^2 + (y_2 - y_1)^2 = (x_3 - 1)^2 + (y_3)^2$$

$$(x_2^2 - 2x_1x_2 + x_1^2) + (y_2^2 - 2y_1y_2 + y_1^2) = (x_3^2 - 2x_3 + 1) + y_3^2 .$$

And after some rearranging we get

$$(x_1^2 + y_1^2) + (x_2^2 + y_2^2) - 2x_1x_2 - 2y_1y_2 = (x_3^2 + y_3^2) - 2x_3 + 1$$

$$1 + 1 - 2(x_1x_2 + y_1y_2) = 1 - 2x_3 + 1$$

$$2 - 2(x_1x_2 + y_1y_2) = 2 - 2x_3$$

$$-2(x_1x_2 + y_1y_2) = -2x_3$$

$$(x_1x_2 + y_1y_2) = x_3$$

or

$$x_3 = x_1x_2 + y_1y_2.$$

And using the values for x_1, x_2, x_3, y_1 and y_2 from above we get

$$\cos(\theta_1 - \theta_2) = \cos\theta_1 \cdot \cos\theta_2 + \sin\theta_1 \cdot \sin\theta_2 \ .$$

When $\theta_1 = \theta_2$ $\cos(\theta_1 - \theta_2) = \cos 0 = 1$

and

$$\cos\theta_1 \cdot \cos\theta_2 + \sin\theta_1 \cdot \sin\theta_2 = (\cos\theta_1)^2 + (\sin\theta_1)^2 = 1$$

which shows that the identity is also true for $\theta_1 = \theta_2$.

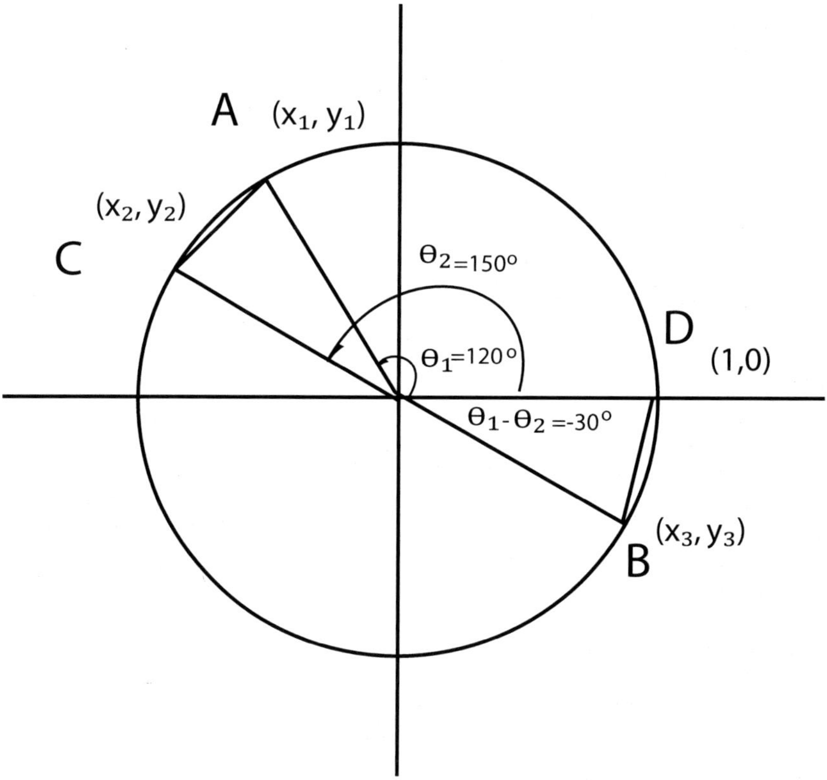

The Figure above shows the case where $0 < \theta_1 < \theta_2 < 2\pi$ and $\theta_1 - \theta_2 = \theta_3$.

The expressions for the distances between the points *A* and *C* and between *B* and *D* are the same as before so that when we equate them we get the same result as before which verifies that $\cos(\theta_1 - \theta_2) = \cos\theta_1 \cdot \cos\theta_2 + \sin\theta_1 \cdot \sin\theta_2$ remains true when $\theta_1 < \theta_2$.

Bibliography

Dennis D. Berkey. *Single Variable Calculus, International Edition.* Saunders College Pub; International Ed edition (1984)

Richard Courant and Herbert Robbins. *What is Mathematics?* Oxford University Press (1979)

Robert Ellis and Denny Gulick. *College Algebra and Trigonometry.* Harcourt Brace Jovanovich, Inc, 1981.

Eric Gossett. *Discrete Math with Proof.* Prentice Hall; United States Ed edition (December 30, 2002)

David Halliday and Robert Resnick. *Fundamentals of Physics, Extended Third Edition.* John Wiley & Sons, Inc., 1988.

Roland E. Larson and Robert P. Hostetler. *Calculus with Analytic Geometry, Third Edition.* D.C. Heath; 3rd ed edition (January 1986)

Eugene D. Nichols, Ralph T. Heimer and E. Henry Garland. *Modern Intermediate Algebra.* Holt, Rinehart and Winston of Canada, Limited, 1970.

James Stewart. *Single Variable Calculus, Early Transcendentals, Second Edition.* Brooks / Cole Publishing Company, 1991.

Index

Please email errors to smf_errata@yahoo.com

Made in United States
North Haven, CT
04 November 2026

21733841R00088